Chemical Weathering of the Silicate Minerals

Chemical Weathering
of the Silicate Minerals

by

F. C. LOUGHNAN

School of Applied Geology
The University of New South Wales
Kensington, N.S.W., Australia

American Elsevier Publishing Company, Inc.
New York 1969

AMERICAN ELSEVIER PUBLISHING COMPANY, INC.
52 Vanderbilt Avenue, New York, N.Y. 10017

ELSEVIER PUBLISHING COMPANY,
335 Jan Van Galenstraat, P.O. Box 211
Amsterdam, The Netherlands

International Standard Book Number 0-444-00043-7
Library of Congress Card Number 69-13273

©American Elsevier Publishing Co., Inc., 1969
Second printing, 1973

Manufactured in the United States of America

CONTENTS

Preface

It is rather surprising that despite the appearance over the past two decades or so of a wealth of literature relating to the chemistry and mineralogy of sediments and soils, textbooks dealing specifically with the mechanisms and products of the chemical weathering of silicate minerals are singularly lacking. The reasons for the void are probably many, not the least being the fear that with the present growth of knowledge, such a book may become obsolete before it is published. It is at the risk of this and other possible criticisms that I have endeavored to fill the gap. Although designed more for graduate students, particularly in the fields of sedimentary petrology, clay mineralogy, pedology, and soil engineering, the book should not prove beyond the capabilities of serious undergraduate students who possess a basic knowledge of chemistry and mineralogy.

The book emphasizes the particular theme that chemical weathering proceeds by reactions that are subject to the same basic laws of chemical equilibria as are normal chemical reactions. Equilibrium is disturbed primarily by meteoric waters that enter the parent mineral and dissolve and effectively remove constituents from the system. Leaching, therefore, is the dominant factor influencing the rate of weathering and the development of specific secondary minerals. Much discussion is also given to the role played by hydrogen ion concentration, or pH, since it is an essential thesis of the book that the environmental pH represents for the most part the effect rather than the cause of chemical weathering. The latter part of the book is devoted to an examination of the mechanisms by which silicate minerals are transformed into new crystalline species and to the association of these secondary minerals with specific weathering environments.

Every effort has been made to present the data in a concise form, for it is realized that in this modern world students and faculty alike have little time to spend grasping more than just the pertinent facts. However, the conciseness is partly alleviated by the liberal use of diagrams, tables, and references.

Acknowledgment and thanks are given to Mr. George Mengyan who freely gave of his time to draft the diagrams.

F. C. Loughnan

Sydney, Australia
January, 1969

LIST OF ILLUSTRATIONS

I. Introduction

Rocks at or near the earth's surface are vulnerable to attack by both physical and chemical processes which may transform them to such an extent that the products bear little resemblance to the original materials. Thus, granite may be reduced to a friable sand of quartz and feldspar grains and basalt may be converted to a soft, white clay such as kaolinite, while limestone may simply disappear in solution leaving a superficial residue of insoluble impurities, such as quartz, clay, and iron oxides, as a reminder of the once massive beds of solid rocks.

Since the breakdown of rocks results from their direct contact with the prevailing atmospheric conditions or weather, the term *weathering* is used to cover these processes. Weathering may be essentially mechanical, that is, the rocks are broken down or disintegrated by physical processes; or it may be chemical, that is, the constituents of the rock undergo chemical changes which lead to their decomposition.

Physical or mechanical weathering involves the fragmentation or comminution of rocks without marked change in chemical composition. Foremost among these processes are stresses induced by the loss of overlying material through erosion, by thermal expansion and contraction resulting from diurnal variations in temperature, by expansion of water upon freezing, by the growth of crystals other than ice, by the disintegrating action of plants and animals, and by erosional forces such as mass wasting, sandblasting, glacial activity, and so on.

On the other hand, chemical weathering involves chemical changes and, contingent upon these, mineralogical changes produced in rocks through their contact with the atmosphere. Minerals comprising the rocks are attacked by water, oxygen, and carbon dioxide of the atmosphere and some of the constituents are removed in solution by the downward percolating meteoric waters. At the same time, the residue recrystallizes to form new mineral phases, the nature of which is determined primarily by the chemical composition of the residue.

As with most natural phenomena, it is difficult to assign a precise definition to weathering. According to Merrill (1906), "The term weathering . . . is applied only to those superficial changes in a rock mass brought about through atmospheric agencies and resulting in a more-or-less complete destruction of the rock as a geological body." He was careful to exclude deeper-seated changes in rocks resulting from hydrothermal and similar processes. Jackson and Sherman (1953) considered that "The term 'weathering of rocks' refers to the changes in degree of consolidation and in compaction which take place in the earth's crust within the sphere of influence of atmospheric and hydrospheric agencies." They subsequently referred to chemical changes but omitted these from the formal definition. Probably the most widely accepted definition is that due to Reiche (1950): "Weathering is the response

1

of materials which were in equilibrium within the lithosphere to conditions at or near its contact with the atmosphere, the hydrosphere and, perhaps still more importantly, the biosphere." Keller (1957) believed this could be improved by deleting "which were in equilibrium" for, as he pointed out, "rocks are in equilibrium momentarily, that is, while the environment in which they were formed persists."

Since we are here concerned only with chemical weathering, perhaps Reiche's definition could be rephrased to read, *Chemical weathering is a process by which atmospheric, hydrospheric, and biologic agencies act upon and react with the mineral constituents of rocks within the zone of influence of the atmosphere, producing relatively more stable, new mineral phases.* The chemical changes taking place are essentially (a) the removal of the more soluble components of the constituent minerals and (b) the simultaneous addition of hydroxyl groups and possibly oxygen and carbon dioxide from the atmosphere. Although the reactions involved are probably similar, it is proposed to exclude from the definition chemical changes which occur in newly deposited sediments on floors of marine and fresh water basins and to which Hummel (1922) applied the term *halmyrolosis.*

The importance of chemical weathering cannot be overstressed for without it the earth's surface would present a lunar landscape and life, as we know it, could not exist. One of its natural consequences is soil formation; the distribution pattern of the great soil groups of the world, with their varying degrees of fertility and productivity, is controlled by the nature of the chemical weathering of the parent rocks. Closely correlated with this distribution pattern is that of man's cultural activities, or, in the words of Keller (1957), "soil is the intermediary mechanism by which chemical weathering has conditioned and will continue to influence man's culture and civilization."

To the geologist bent upon unraveling the past, an understanding of the processes and products of chemical weathering is essential for, as Pettijohn (1957) has pointed out, one of the principal objectives of the geologist is to determine from the examination of the composition and maturity of sediments the climate, relief, and nature of the rocks that comprised the source area. In addition, chemical weathering has been the mechanism for the concentration of many valuable ores and industrial minerals including bauxites, iron ores, clays, and building materials, upon which modern civilization is becoming increasingly dependent.

Moreover, as shown in a later chapter, there exists a distinct correlation between the degree of chemical weathering and the type of clay minerals developed in the resulting soils. Consequently, since the clay minerals exhibit a considerable variation in rheological properties, such as plasticity, shear strength, and so on, the nature of the chemical weathering is of some concern to the engineer who is required to build structures and roads on, through, or possibly out of these soil materials.

Since chemical weathering is essentially a process of mineral transformation, it seems appropriate, before we proceed to an examination of the mechanisms involved, that we give at least brief consideration to the more important minerals participating in these transformation reactions. Consequently, Chapter II is devoted to a description of the properties and crystalline structures of the silicate minerals commonly encountered in rocks, and their weathered derivative products. Subdivision is made into *primary minerals*, that is, those of magmatic, hydrothermal, and metamorphic origin, and *secondary minerals*, that is, those formed as a result of weathering. Such a subdivision is not strictly valid, for clay minerals, the common weathered derivative products, may also originate under hydrothermal conditions while many of the so-called "primary minerals" such as feldspar, quartz, and tourmaline are known to develop in sedimentary environments. Nevertheless, from a descriptive viewpoint, the groupings are useful.

The chemistry of weathering is discussed in Chapter III. The mechanism of breakdown of the parent mineral structures is examined and the roles played by the hydrogen ion concentration (pH), redox potential (Eh), fixation, chelation, and degree of leaching are assessed. Finally, consideration is given to the influence exerted by the parent mineral structures on the rate of chemical weathering.

In Chapter IV a correlation is made between the chemical factors, particularly the degree of leaching and the redox potential, on the one hand, and the environmental factors of climate, topography, and permeability of the parent rock, on the other.

Examples illustrating the chemical and mineralogical transformations which accompany the weathering of rocks are given in Chapter V. For convenience, the parent materials have been grouped into basic crystalline, acid crystalline, alkaline, and argillaceous rocks. Much of the data for this section have been obtained from published literature.

Chemical weathering in relation to soil formation is discussed in Chapter VI. The principles of *pedology* (soil science) are outlined and brief descriptions of the profile development and mineralogy of some of the great soil groups of the world are given.

REFERENCES

Hummel, K. (1922), Die Entstehung eisenreicher Gesteine durch Halmyrolse, *Geol. Rundschau* **13**, 40–81.

Jackson, M. L., and G. D. Sherman (1953), Chemical weathering of minerals in soils, *Advan. Agron.* **5**, 219–318.

Keller, W. D. (1957), *The Principles of Chemical Weathering*, 2nd ed. Lucas Bros., Columbia, Missouri.

Merrill, G. P. (1906), *Rocks, Rock-Weathering and Soils*. MacMillan, New York.

Pettijohn, F. J. (1957), *Sedimentary Rocks*, 2nd ed. Harper, New York.

Reiche, P. (1950), A survey of weathering processes and products, *Univ. New Mexico Publ. in Geology* **3**, Rvd. Ed.

II. Structures and Properties
of Some of the Primary and Secondary Minerals
Involved in Weathering Reactions

Silicate minerals possess a crystalline structure or lattice in which cations and anions are tightly packed. The spatial configuration is such that each cation is surrounded by and bonded to ions of opposite charge. The bonding is both ionic and covalent and the proportion of each in any particular linkage is dependent on the nature of the cations and anions involved or, more specifically, the difference in electronegativity between the two (Pauling, 1940). The number of anions which may surround and bond to a given cation is limited and depends on the radii and valencies of both the cation and anion. This value is termed the *coordination number*. In silicate structures, oxygen is by far the most important anion and, to simplify matters, only the coordination numbers for the different cations with respect to oxygen need be considered. The effective ionic radii in Ångström units ($1Å = 10^{-8}$ cm), the coordination numbers, and the proportion of ionic bonding for the various cations with oxygen are given in Table 1. It will be observed that the number of cation-oxygen bonds increases with the radii of the cations from 4 in the case of silicon to 12 or perhaps 14 with potassium, rubidium, and cesium. Fourfold coordination is termed *tetrahedral*; 6-fold, *octahedral*; 8-fold, *cubic*; and 12-fold, *dodecahedral*.

TABLE 1

Coordination Numbers and Nature of the Bonds for Some Common Cations with Oxygen
($= 1.40$ Å)

Ion	Ionic radius (Å)	Radius ratio, cation/oxygen	Coordination number	% Ionic bond
Si^{4+}	0.41	0.29	4	51
Al^{3+}	0.50	0.36	6, 4	63
Li^+	0.60	0.43	6	79
Fe^{3+}	0.64	0.46	6	51
Mg^{++}	0.65	0.46	6	74
Ti^{4+}	0.68	0.49	6	63
Fe^{++}	0.76	0.54	6	72
Zr^{4+}	0.80	0.57	6, 8	67
Na^+	0.95	0.68	8	82
Ca^{++}	0.99	0.71	8	79
Sr^{++}	1.13	0.81	8	79
K^+	1.33	0.95	8, 12, (14)	84
Rb^+	1.48	1.06	12, (14)	84
Cs^+	1.69	1.21	12, (14)	86

4

The concept of coordination numbers has a bearing on another important phenomenon, namely, *isomorphous substitution*. Cations of similar coordination number may replace or proxy for one another, either completely or to a limited degree, in silicate structures provided that

    (a) the radii of the two ions are not appreciably different,

    (b) the difference in charge is not greater than one valency unit, and

    (c) the structure is suitable.

Thus, ferrous and magnesium ions have similar radii and in many silicate structures, olivine for example, a complete isomorphous series exists between ferrous and magnesium end members. On the other hand, although aluminum may exist in fourfold coordination, its replacement of silicon is restricted to a 1 : 1 ratio as in the plagioclase feldspars, while in certain minerals such as kaolinite, it is inhibited from entering the tetrahedral part of the structure altogether.

According to Pauling (1940), a stable configuration is possible only if the sums of the positive and negative charges equate, that is, a mineral must be in electrical neutrality. Consequently, if a cation of lower positive charge, Al^{3+} for instance, replaces one of higher positive charge, such as Si^{4+}, another cation must be introduced into the lattice to balance the deficit. In the case of the plagioclase feldspars for example, albite ($NaAlSi_3O_8$) has one in every four of its silicons replaced by aluminum and Na^+ balances the structure. Anorthite ($CaAl_2Si_2O_8$), on the other hand, has two silicons replaced by aluminums and a divalent cation (calcium) must be introduced into the structure to achieve electrical neutrality.

Another important concept arising out of the study of the size and charge of ions is that of *ionic potential*. Cartledge (1928) was the first to propose the concept and defined it as the ratio of the charge in valency units to the ionic radius in Ångström units. The polarizing power of a cation with respect to oxygen, that is, the ability of the cation to distort the electron shells of bonded oxygens, increases with the ionic potential and, since the stability of the bond decreases with increased polarization of the oxygen, ionic potential has an

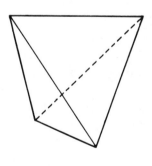
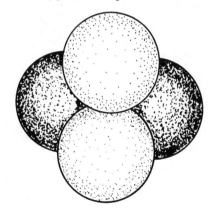

Figure 1.   The silica tetrahedron.

important bearing on the stability of cation-oxygen linkages. The influence of ionic potential on silicate structures was stressed by Ramberg (1954) when he concluded, "Not only is the question whether an element can form silicates or not controlled by the field strength (i.e. the ionic potential) around the cation, but so is the maximum degree of polymerization which silicates can have."

All silicate minerals have, as their basic structural unit, the silica tetrahedron (Fig. 1) in which the silicon ion (possibly in part replaced by Al^{3+}) is equi- or approximately equidistant from four oxygens, the centers of the latter forming the corners of the tetrahedron. Polymerization of the tetrahedra is possible and the classification of the silicate minerals is based on the degree of polymerization.

THE PRIMARY MINERALS

1. Nesosilicates (SiO₄ Group)

In contrast to the remaining silicate groups, polymerized tetrahedra are absent in the nesosilicates. Instead, the structure consists of independent SiO_4

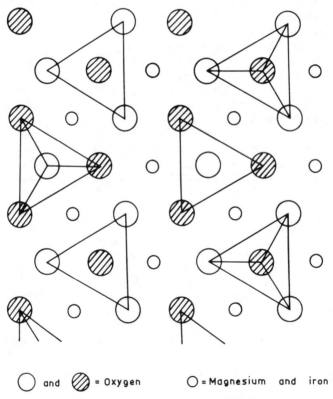

◯ and ⬯ = Oxygen   ◯ = Magnesium and iron

Figure 2.   The structure of olivine (after Bragg, 1937).

units bonded together by cations. Each oxygen of the tetrahedra has one of its negative charges satisfied by the silicon ion and the other by bonding cations which can be in four-, six-, or eightfold coordination. Aluminum substitution for silicon within the teterahedra is rare.

The *olivines*, which are common representatives of the group, have discrete silica tetrahedra linked by magnesium (*forsterite*, Mg_2SiO_4) and ferrous (*fayalite*, Fe_2SiO_4) ions in octahedral coordination (Fig. 2) and a complete isomorphous series exists between the two end members. Cleavage is poorly developed in the olivines.

In *zircon* the zirconium ion is in eightfold coordination and in *kyanite* aluminum is in sixfold coordination. Other more complex members of the group include *garnets*, with a general formula $R_3^{++}R_2^{3+}(SiO_4)_3$; *sphene*, $CaTiSiO_5$: *sillimanite*, Al_2SiO_5; *andalusite*, Al_2SiO_5; and *topaz*, $Al_2SiO_4(F,OH)_2$.

From a chemical weathering viewpoint, the nesosilicates present an interesting group for although the olivines are among the least stable of all silicates, zircon and garnets are remarkably resistant to breakdown and generally persist in the weathering residue. Marshall (1964) has attributed this considerable disparity in weathering stability to the presence of sparingly soluble ions in zircon (Zr) and the garnets (Al^{3+} and Fe^{3+}).

2. Sorosilicates (Si_2O_7 Group)

The sorosilicate structure consists of a polymerization of two silica tetrahedra with a single Si–O–Si bond. The tetrahedra are rotated through 180° with respect to each other. Representatives of the group are not common but include *gehlenite*, $Ca_2Al(AlSi)_2O_7$, and *äkermanite*, $Ca_2MgSi_2O_7$.

3. Cyclosilicates (Si_6O_{18} Group)

In the cyclosilicates the tetrahedra are arranged in hexagonal rings with each SiO_4 unit sharing two of its oxygens with adjacent units (Fig. 3). The rings

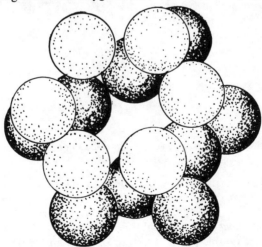

Figure 3.   Arrangement of silica tetrahedra in the cyclosilicates.

are stacked above one another in the c crystallographic direction and the hexagonal columns so created are bonded together by cations. In *beryl*, $Be_3Al_2Si_6O_{18}$, the bonding cations are aluminum in octahedral coordination and beryllium in tetrahedral coordination. Although alkali ions do not appear essential to the structure of beryl, frequently they are present in small amounts and occupy the cavity within the hexagonal framework. The structure of *cordierite*, $Al_3Mg_2(Si_5Al)O_{18}$, is similar to beryl. In *tourmaline*, $(Na, Ca)(Li, Al)_3(Fe, Mn, Al)_6(BO_3)_3(Si_6O_{18})$, however, the configuration of the intercolumnar areas is more complex with aluminum, magnesium, and iron in sixfold coordination and boron in threefold coordination.

The common cyclosilicates are relatively resistant to attack by weathering solutions and tend to persist as detrital grains in secondary rocks.

4. Inosilicates (Si_2O_6 and Si_4O_{11} Groups)

The inosilicates comprise two important groups of rock-forming minerals, the *pyroxenes* and the *amphiboles*.

In the pyroxene or Si_2O_6 structure, the silica tetrahedra point in the one direction and are arranged into single chains which parallel the c crystallographic axis. Each tetrahedral unit shares two of its oxygens with adjacent units (Fig. 4) and the chains are linked together laterally through the unshared oxygen, by cations either in octahedral or cubic coordination. A limited

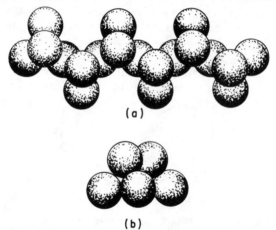

(a)

(b)

Figure 4.   The single-chain arrangement of tetrahedra in the pyroxenes. (a) Parallel to the c axis; (b) normal to the c axis.

degree of aluminum substitution for silicon in the tetrahedral units is evident in some of the more complex members of the group. The composition of the common pyroxenes is shown in Fig. 5. Sodium-bearing pyroxenes are also known and include *aegerine*, $NaFeSi_2O_6$, and *jadeite*, $NaAlSi_2O_6$.

The Si_4O_{11} or double-chain structure of the amphiboles is formed by a polymerization of two pyroxene chains (Fig. 6). However, unlike the pyro-

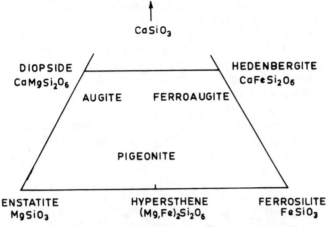

Figure 5.   Triangular diagram showing the composition of some pyroxenes.

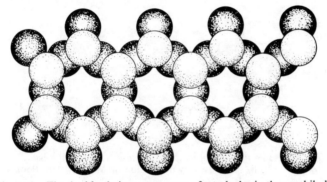

Figure 6.   The double-chain arrangement of tetrahedra in the amphiboles.

xenes, the amphiboles possess hydroxyl ions which occupy the cavities created by the linking of the two chains. The double chains are bonded together by Ca, Mg, Al, Fe, and Na ions. The composition of some of the more important amphiboles is given in Table 2.

TABLE 2
Composition of Some Common Amphiboles

Orthorhombic	
Anthophyllite	$(MgFe)_7Si_8O_{22}(OH)_2$
Monoclinic	
Tremolite	$Ca_2Mg_5Si_8O_{22}(OH)_2$
Actinolite	$Ca_2(Mg, Fe)_5Si_8O_{22}(OH)_2$
Hornblende	$(Ca, Na)_2(Mg, Fe, Al)_5(Al, Si)_8O_{22}(OH)_2$
Riebeckite	$Na_2Fe_4Si_8O_{22}(OH)_2$
Glaucophane	$Na_2(Mg, Al, Fe)_5Si_8O_{22}(OH)_2$

The inosilicates cleave readily in two directions parallel to the lengths of the silica chains. In the pyroxenes, these cleavages intersect at approximately 90° but in the amphiboles they make angles of 56° and 124° with each other. These planes of weakness within the crystal structure facilitate easy access of water to the cations bonding the chains together and, where the cations are readily soluble, breakdown of the structure is rapid. Consequently inosilicates rarely survive the weathering environment.

5. Phyllosilicates (Si_4O_{10} Group)

Further polymerization of the silica tetrahedra results in a sheetlike hexagonal or pseudohexagonal network (Fig. 7) in which the apices of the tetrahedra point in the one direction. Hydroxyl ions occur in the plane of the

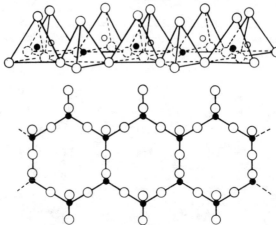

Figure 7.   The hexagonal network of silica tetrahedra in the phyllosilicates. (from *Clay Mineralogy* by R. E. Grim, 1953; used by permission of McGraw-Hill Book Company).

apical oxygens and, in this respect, the structure is similar to that of the amphiboles. The primary phyllosilicates contain two such sheets inverted to each other and bonded together through the unshared oxygens by cations such as Al^{3+}, Mg^{++}, Fe^{3+}, and Fe^{++} in octahedral coordination. The structure, therefore, consists of a close-packed octahedral sheet sandwiched between two polymerized silica networks (Fig. 8).

Where the bonding cations are trivalent, only two thirds of the octahedral cation sites are filled and these phyllosilicates are termed *dioctahedral*. The *trioctahedral* phyllosilicates, on the other hand, have all cation sites filled with divalent ions. *Pyrophyllite*, $Al_2Si_4O_{10}(OH)_2$, has the basic structure of the dioctahedral, and *talc*, $Mg_3Si_4O_{10}(OH)_2$, of the trioctahedral phyllosilicates. In these minerals the unit layers, each composed of one octahedral and two tetrahedral sheets, have a thickness of 9.1–9.4 Å. The layers are stacked on one another and the interlayer bonding is by Van der Waals forces. The *a* and *b* crystallographic axes lie in the plane of the layers and the *c* axis is normal to that plane.

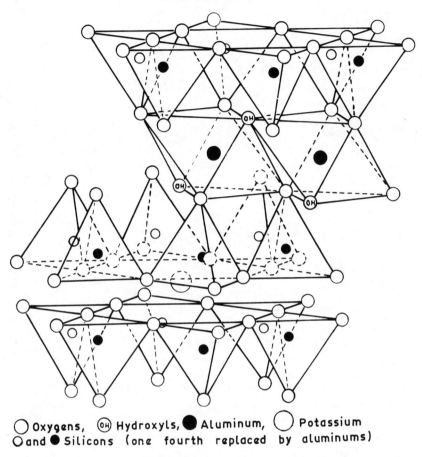

○ Oxygens, ⓞ Hydroxyls, ● Aluminum, ○ Potassium
○ and ● Silicons (one fourth replaced by aluminums)

Figure 8.   The muscovite structure.

In the micas, one in four of the silicons of the tetrahedral sheets is replaced by aluminum and the charge is neutralized by the introduction of potassium (more rarely sodium) into the structure. The potassium ions are located between the layers and assist in bonding the layers together. Nevertheless, the bonding is weak and a well-developed cleavage parallels the layers. The potassium ions fit partly into the perforations in the silica sheets of adjacent layers but, because these perforations are distorted from hexagonal symmetry (Radoslovich, 1960), the potassium projects above the level of the tetrahedra and the layers are separated by the order of 0.7–0.8 Å.. The total thickness of the layers is thereby increased to 10Å.

   Muscovite, $KAl_2(AlSi_3)O_{10}(OH)_2$, is the mica analog of pyrophyllite and despite the apparent vulnerability of the potassium ions to solution, it is one of the most resistant of the silicate minerals to breakdown in the weathering

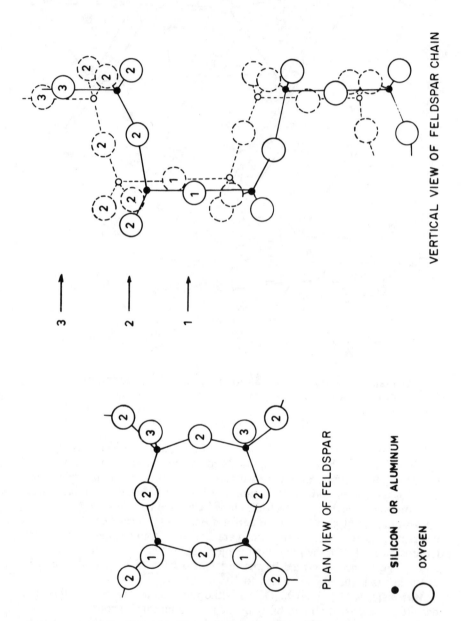

Figure 9.   The chain structure of the feldspars.

environment. Up to the present no satisfactory explanation has been advanced to account for the remarkable stability of muscovite.

On the other hand, the trioctahedral micas, *biotite*, $K(MgFe)_3(AlSi_3)O_{10}$ $(OH)_2$, and *phlogopite*, $KMg_3(AlSi_3)O_{10}(OH)_2$, weather rapidly. Attack commences along the cleavages and results in solution of potassium ions. If the weathering is not particularly aggressive, the residual layers may become stabilized by the absorption of magnesium, aluminum, or perhaps ferrous ions, but under prolonged attack ions are leached from the octahedral sheet and a complete breakdown of the structure ensues.

6. Tectosilicates (Si_4O_8)

The most complex silicate structure is the three-dimensional framework of tetrahedra in which all oxygens are shared by silicons. The group includes the most important primary silicate minerals, quartz and feldspars, as well as the feldspathoids and the zeolites.

Quartz. Although silica can crystallize in a number of polymorphic forms, namely quartz, tridymite, cristobalite, coesite, and stishovite, with various low (α) and high (β) temperature modifications, α quartz is the only naturally occurring representative of any real significance. The structure consists of a spiral arrangement of tetrahedra, which may twist to either the left or right, thereby giving rise to two enantiomorphous forms. At a temperature of 573 °C α quartz inverts to the β form and at still higher temperatures, conversion to tridymite and cristobalite takes place. According to Krauskopf (1959), quartz has a solubility of about 7–14 ppm near neutrality and, consequently, it is one of the most resistant minerals to chemical attack. Nevertheless, exposure to intense leaching over prolonged periods of time may result in considerable depletion of the mineral from the weathering environment (Loughnan and Bayliss, 1961).

Feldspars. In the feldspars, partial replacement of silicon by aluminum in tetrahedral coordination occurs to the maximum extent of 1 : 1 and the structure is balanced by the introduction of alkali and alkaline earth ions. The silica and alumina tetrahedra are arranged in rings which form continuous chains (Fig. 9) paralleling the *a* axis, and the larger alkali or alkaline earth ions fit into the cavities created by the linkage of the chains. Three distinct feldspar families are recognizable: (a) the potash feldspars including *sanidine*, *adularia*, and *microcline* with the formula $KAlSi_3O_8$, (b) the *plagioclases*, which form a complete isomorphous series between *albite* ($NaAlSi_3O_8$) and *anorthite* ($CaAl_2Si_2O_8$), and (c) the rare barium feldspar, *celsian* ($BaAl_2Si_2O_8$).

Although a great deal of experimental work has been carried out on the weathering of the feldspars (Marshall, 1964), the mechanisms are imperfectly understood. Observations on natural weathering environments have established that anorthite is the least stable of the feldspars and chemical breakdown is relatively rapid. Kaolinite is generally the final product, but aluminous montmorillonite and even chlorite may develop as intermediate phases

(Craig and Loughnan, 1964). The potash and soda feldspars, on the other hand, offer greater resistance to chemical attack and these minerals tend to persist as detrital grains in sedimentary rocks. Illite, kaolinite, and halloysite are the common alteration products.

Feldspathoids. The feldspathoids are a group of minerals that are closely related to the feldspars but contain a lower silica : alkali ratio. They tend to take the place of the feldspars in silica-deficient igneous rocks. *Nepheline* ($NaAlSiO_4$) is the more prevalent of the sodium feldspathoids and *leucite* ($KAlSi_2O_6$) that of the potassium-bearing feldspathoids. Both minerals are very unstable under weathering conditions.

Zeolites. The zeolites form a relatively large group of hydrous alumino-silicates, which contain alkalies or alkaline earths or both and which possess a three-dimensional framework of tetrahedra. They characteristically occur as cavity fillings in basic lavas and as hydrothermal alteration products of primary minerals, particularly calcic feldspars. However, zeolites of diagenetic and low-grade metamorphic origin are also quite common. *Analcime* (analcite), $NaAlSi_2O_5(OH)_2$, *natrolite*, $Na_2Al_2Si_3O_8(OH)_4$, *heulandite*, (Ca Na$_2$)(Al$_2$Si$_7$O$_{12}$)(OH)$_{12}$, and *stilbite*, (CaNa$_2$K$_2$)(Al$_2$Si$_7$O$_{11}$)(OH)$_{14}$, are probably the more abundant members of the group. The zeolites are particularly vulnerable to destruction under weathering conditions.

THE SECONDARY MINERALS

The term *secondary mineral* is used here to include those which have crystallized *in situ* from atoms and ions not removed by the weathering processes, that is, the residual derivative minerals. The clay minerals and the oxides and hydroxides of the stable elements, namely aluminum, ferric iron, and titanium, are the most important members.

1. *The Clay Minerals*

The development and improvement of analytical techniques such as X-ray diffraction, differential thermometry, electron microscopy, and infrared spectroscopy in the last three to four decades have greatly enhanced our knowledge and understanding of the nature and property of clays.

Early concepts of clays and clay rocks envisaged the predominant components as "amorphous substances" which bonded together scattered crystalline grains of quartz, feldspar, and so on. It is now recognized that these supposedly "amorphous substances" in reality are fine, generally flake shaped crystalline particles that can be classified according to their structure. Considerable research has revealed that the properties of the individual clay minerals are primarily a function of their crystalline structure.

In Table 3, the common clay minerals occurring in soils and clay rocks are arranged in accordance with accepted practice into five groups while a sixth, termed *mixed-layered minerals*, includes those with more than one type of layer present, a phenomenon which appears peculiar to the clay minerals.

TABLE 3

Classification of the Clay Minerals

Kaolin Minerals. Nonexpanding. One Tetrahedral Sheet and One Octahedral Sheet.

Dioctahedral series

Kaolinite	$Al_2Si_2O_5(OH)_4$
Disordered kaolinite	$Al_2Si_2O_5(OH)_4$
Dickite	$Al_2Si_2O_5(OH)_4$
Nacrite	$Al_2Si_2O_5(OH)_4$
Halloysite (endellite)	$Al_2Si_2O_5(OH)_4 \cdot 2H_2O$
Metahalloysite (halloysite)	$Al_2Si_2O_5(OH)_4$

Trioctahedral series

Antigorite	$Mg_3Si_2O_5(OH)_4$
Chrysotile	$Mg_3Si_2O_5(OH)_4$
Greenalite	$Fe_3^{++}Si_2O_5(OH)_4$
Cronstedite	$(Fe_2^{++}Fe^{3+})(Fe^{3+}Si)O_5(OH)_4$
Amesite	$Mg_2Al(AlSi)O_5(OH)_4$

Mica Minerals. Nonexpanding. Two Tetrahedral Sheets and One Octahedral Sheet.

Dioctahedral series

Illite (3T)
Illite (2M)
Illite (1M)
Illite (1Md)
$KAl_2(AlSi_3)O_{10}(OH)_2$
With K less than unity and Mg^{++}, Fe^{++}, Fe^{3+} partly replacing Al^{3+}

Glauconite	$K(Fe^{3+}Fe^{++}AlMg)_2(AlSi_3)O_{10}(OH)_2$

Trioctahedral series

Illites rich in magnesium and ferrous iron.

Montmorillonite Minerals. Expanding. Two Tetrahedral Sheets and One Octahedral Sheet.

Dioctahedral series

Beidellite	$Al_4(Si_{7.33}Al_{0.67})O_{20}(OH)_4...M_{0.67}^+$
Montmorillonite	$(Al_{3.33}Mg_{0.67})Si_8O_{20}(OH)_4...M_{0.67}^+$
Nontronite	$Fe^{3+}{}_4(Si_{7.33}Al_{0.67})O_{20}(OH)_4...M_{0.67}^+$

Trioctahedral series

Saponite	$Mg_6(Si_{7.33}Al_{0.67})O_{20}(OH)_4...M_{0.67}^+$
Hectorite	$(Mg_{5.33}Li_{0.67})Si_8O_{20}(OH)_4...M_{0.67}^+$

Chlorite-Vermiculite Minerals. Nonexpanding. Two Tetrahedral Sheets and Two Octahedral Sheets.

Dioctahedral series

Clay minerals with dioctahedral chlorite- and vermiculite-like structures have been recognized but structural formulas are unknown.

Trioctahedral series

Clay minerals with structures related to chlorite $(MgFe)_{6-x}(AlFe)_x(Si_{4-x}Al_x)O_{10}(OH)_8$ and to vermiculite $Mg_3(AlSi)_4O_{10}(OH)_2.Mg_{0.45}.4.5H_2O$ are widespread in sedimentary rocks and some soils.

Fibrous Clay Minerals

Palygorskite (attapulgite)	$(AlMgFe)_5(AlSi_7)O_{20}(OH)_2 \cdot 4H_2O \cdot 4H_2O$
Sepiolite	$Mg_9Si_{12}O_{30}(OH)_6 \cdot 4H_2O \cdot 6H_2O$

Some disparity in status exists between the groups for, as Weaver (1958) has pointed out, illites, montmorillonites, and chlorites have similar structures, the essential differences between them being the amount of the positive charge deficit and the nature of the sorbed cations. Kaolins, on the other hand, have a distinctive structure and their alteration from and possibly to illite, montmorillonite, and chlorite involves major structural adjustments.

In considering weathering and authigenic development of the clay minerals, it is important that this disparity in status of the groups be borne in mind.

a. *The Kaolin Group.* Minerals of the kaolin groups have a layered structure similar to the phyllosilicates but differ from the latter in containing only one tetrahedral and one octahedral sheet per unit layer (Fig. 10). The tetrahedral

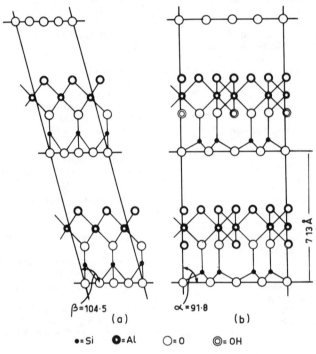

$\bullet = Si$ $\bullet = Al$ $\bigcirc = O$ $\circledcirc = OH$

Figure 10.   The structure of kaolinite (after Brindley, 1951).

sheet is 4.94Å high and the octahedral sheet 5.04Å, while the distance separating the center planes of the two sheets is approximately 2.1Å. Consequently, the thickness of the unit layer is of the order of 7.1Å (made up of half the tetrahedral sheet, 2.47Å plus half the octahedral sheet, 2.52Å, and the intersheet distance, 2.1Å.)

The crystal structure consists of unit layers stacked on one another in the c crystallographic direction and held together firmly by hydrogen bonding between the hydroxyl ions of the octahedral sheet of one layer and the oxygens of the tetrahedral sheet of an adjacent layer. Polymorphism in the kaolin group results from the displacement of adjacent layers relative to one another in the a-b axial plane. The shifts are possible in both the a and b crystallographic directions and the polymorphic forms may be represented by the symbols $na_{0/6}$ and $mb_{0/6}$ where n and m are integers and a_0 and b_0 are the unit cell axial dimensions.

In well-crystallized *kaolinite* $n=2$ *and* $m=0$ but disordered forms of the mineral having random stacking, particularly in the b direction, are common. *Nacrite* and *dickite*, the former with six and the latter with two layers per unit cell, have a greater degree of crystallinity than kaolinite. However, neither has been recorded as a product of weathering.

Kaolinite, disordered kaolinite, nacrite, and dickite are dioctahedral kaolins in which aluminum is the only cation in octahedral coordination and isomorphous substitution in either the tetrahedral or octahedral sheet does not occur or is restricted to a very limited degree.

Halloysite is also a member of the dioctahedral, aluminous kaolins but, unlike the preceding, it exists in two states of hydration. The fully hydrated form, to which the terms "halloysite", "hydrated halloysite," and "endellite" have been applied, has a structure similar to disordered kaolinite but contains in addition a single sheet of oriented water molecules between the layers (Fig. 11). The presence of this additional water increases the basal spacing

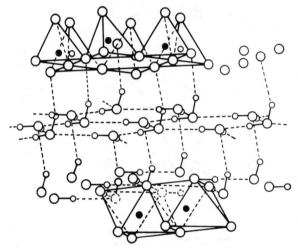

Figure 11.   Configuration of interlayer water in hydrated halloysite (after Hendricks and Jefferson, 1938).

from 7.2Å to 10Å. Even greater expansion is possible with the addition of certain polar, organic liquids such as glycol. On drying in a vacuum or at 100°C, the water is lost irreversibly and the structure collapses to 7.2Å. The terms *metahalloysite* and *halloysite* have been applied to the collapsed form. All intermediate stages between the two limits of hydration occur in nature and may be produced under carefully controlled conditions in the laboratory. Electron microscopy has revealed that most, but certainly not all, halloysites occur as rolled, tubelike crystals which contrast with the platy characteristics of other dioctahedral kaolin minerals.

Trioctahedral kaolins in which divalent cations, principally ferrous iron and magnesium, have completely replaced aluminum in octahedral coordination are also known and include the serpentine minerals, chrysotile and antigorite. However, because of the mobility of the divalent cations, trioctahedral kaolins are particularly unstable and tend to be destroyed rapidly in the weathering environment. Trioctahedral kaolins, such as *greenalite*, *cronstedite*, *amesite*, and probably some of the *chamosites*, are relatively common as authigenic minerals in ferrous- and magnesium rich sediments and often considerable difficulty is experienced in distinguishing them from dioctahedral kaolins and clay-chlorites.

b. *The Illite Group*. The term *illite* was introduced by Grim, Bray, and Bradley (1937) to include the clay minerals which possess affinities with the micas of the igneous and metamorphic rocks. Like their well-crystallized counterparts, the illites have a unit layer composed of an octahedral sheet sandwiched between two tetrahedral sheets (Fig. 12). Each tetrahedral sheet

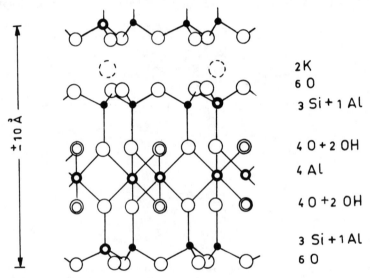

$\pm 10 \text{Å}$

2 K
6 O
3 Si + 1 Al

4 O + 2 OH
4 Al
4 O + 2 OH

3 Si + 1 Al
6 O

Figure 12.   The illite structure (from *Clay Mineralogy* by R. E. Grim, 1953; used by permission of McGraw-Hill Book Company).

has a thickness of 4.94Å and the distance between the center planes of the tetrahedral and octahedral sheets is 2·1Å. Consequently, the illite unit layer has a thickness of a little over 9Å ($\frac{1}{2} \times 4.94 + 2.1 + 2.1 + \frac{1}{2} \times 4.94$). However, potassium ions occur between the sheets and these cause a separation of the layers by a further 0.7–0.8Å, making the total thickness for the illite layer approximately 10Å. Polymorphism results from differences in the stacking of the layers and Levinson (1955) has identified four structures corresponding to a three-layer trigonal illite (3T), a two-layer monoclinic illite (2M), a one-layer monoclinic illite (1M), and a one-layer disordered monoclinic illite (1Md).

According to Grim (1953), illites differ from true micas in several ways:

(i) Aluminum replacement of silicon in the tetrahedral sheets is close to 1 : 6, whereas in the true micas it is 1 : 4.

(ii) Calcium, magnesium, and hydrogen ions partly replace potassium in the interlayer position and water generally is present.

(iii) The stacking of the layers is more random.

(iv) They generally occur as very fine particles whereas the well-crystallized micas are relatively coarse.

Both dioctahedral and trioctahedral illites are known but, although every gradation between the two groups is theoretically possible, observations on a considerable number of illites have shown that there is a marked tendency for them to be distinctly one or the other.

The illites are the most abundant clay minerals in sediments and are also quite common in soils, particularly those in which leaching has not been excessive. Generally the illites of soils and recent sediments are degraded in that there is a deficiency of interlayer potassium ions and the structures have a tendency to expand in the presence of water in much the same manner as the montmorillonites.

According to Hendricks and Ross (1941), *glauconite* is a dioctahedral illite in which ferric ions predominate in the octahedral positions although aluminum, magnesium, and ferrous ions are almost invariably present. Burst (1958), however, investigated a wide variety of glauconites and concluded that "the term 'glauconite' is currently being used with a dual connotation. Originally coined as a description for a blue-green micaceous mineral, this word is now used as a morphological term for small, spherical, green, earthy pellets. By X-ray these pellets can be grouped in four morphological classifications, only one of which has the diffraction properties usually attributed to the mineral glauconite. These differences are not necessarily reflected in the size, shape or color of the pellets."

c. *The Montmorillonite Group*. Minerals of the montmorillonite group occur as extremely fine particles and single crystals of sufficient size for X-ray examination are unknown. Consequently, many uncertainties as to the true nature of the crystal structure still exist. However, mainly through the work of Hofmann, Endell, and Wilm (1933), Marshall (1935), Maegdefrau and Hofmann (1937), Hendricks (1942), and Ross and Hendricks (1945), it is generally recognized that the montmorillonite structure is analogous to pyrophyllite (or talc) but differs from the latter in that isomorphous substitution is widespread in both the tetrahedral and octahedral sheets and the layers lack electrical neutrality. In the dioctahedral montmorillonites, the principal replacements are Mg^{++} for Al^{3+} and Al^{3+} for Si^{4+}, and the negative charge on the lattice created by these substitutions tends to be balanced by the sorption of cations at exposed surfaces and between the layers (Fig.13). Water and other polar liquids, either arranged in definite groups about the counterbalancing cations or adsorbed directly on to the surfaces, may enter

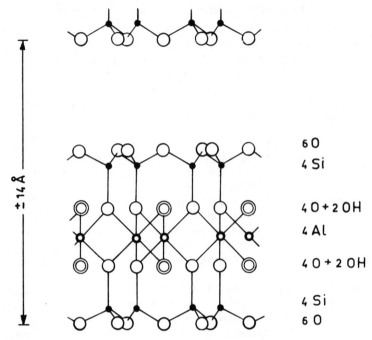

Figure 13. The montmorillonite structure.

between the layers and cause expansion of the structure in a direction normal to the plane of the layers (i.e., in the c crystallographic direction). Montmorillonites dried at temperatures in excess of 200°C have an interlayer distance ranging between 9.1 and 10.0Å (cf. pyrophyllite, 9.1Å, and talc, 9.3Å), depending on the nature of the adsorbed cation, but on saturation with water the structure may expand to 10 times this value.

Since both illite and montmorillonite have a substituted pyrophyllite (or talc) structure, it would seem reasonable to assume that the essential difference between the two mineral groups lies in the presence of neutralizing potassium ions in the illites. That this is not true, however, is shown by the difference in behavior between illite stripped of its potassium ions and montmorillonite. Degraded illite (i.e., illite stripped or partially stripped of K^+) expands in a manner similar to montmorillonite but, on the addition of sufficient K^+, the structure collapses irreversibly to 10Å and illite is regenerated. Montmorillonite, on the other hand, if saturated with K^+, undergoes only a partial collapse and much expandable material remains.

According to Grim there are two important differences between the structures of illite and montmorillonite.

(i) "The charge deficiency, due to substitution, per unit layer is about 1.30 to 1.50 for illite compared with about 0.65 for montmorillonite.

(ii) "The seat of this charge deficiency in illite is largely in the silica sheet and therefore close to the surface of the unit layers whereas in montmorillonites it is frequently, perhaps chiefly, in the octahedral sheet at the center of the unit layer."[1]

Within the montmorillonite group several mineral species have been recognized and classification (Table 3) is based primarily on chemical composition. MacEwan (1961) has considered that there is "very limited mutual solubility" between the dioctahedral and trioctahedral series but within each series all gradations are possible.

Discrete montmorillonite, that is, montmorillonite not interlayered with other clay structures, is not particularly abundant in sediments and soils. However, it forms the essential constituent of bentonite, a clay derived from the alteration of volcanic ash *in situ* (Ross and Shannon, 1926), and occurs below the water table in some soils formed from volcanic and other igneous rocks, greywackes, and high-grade metamorphic products.

d. *The Chlorite Group.* The structure of the well-crystallized chlorites consists of talclike layers, separated from each other by fully hydrated, magnesium-rich, octahedral units, termed "brucite layers". Since the talc and brucite layers have thicknesses of the order of 9Å and 5Å respectively (see Fig. 14), the height of the unit structure approximates 14Å. Isomorphous substitution occurs in both the talc and brucite layers. In the former there is a positive charge deficiency which is just balanced by an excess positive charge in the latter and the entire structure is neutral.

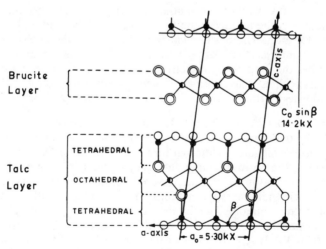

Figure 14.   Projection of the chlorite structure on the *ac* plane.

[1] From *Applied Clay Mineralogy* by R. E. Grim (1962); used with permission of McGraw-Hill Book Company.

brucite layer   talc layer
$$[(Mg_2Al)(OH)_6]^{+1}   [Mg_3(AlSi_3)O_{10}(OH)_2]^{-1}$$

The two parts of the structure are held together by electrostatic charges and by hydrogen bonding through the hydroxyl ions of the brucite layer being in juxtaposition with oxygens of the talc layer. Within the group, the individual species are differentiated on the bases of chemical composition and manner of stacking of the layers.

Degradation of chlorite generally proceeds with the preferential removal of the brucite layers resulting in expanding structures, resembling montmorillonite, as intermediate weathered phases. These expanding structures have a relatively high charge deficiency and, in consequence, behave in a manner similar to degraded illite.

The chlorites of sediments and soils, the so-called "clay chlorites," invariably occur admixed with other clay minerals and much has yet to be learned concerning their structures and properties. Compared with the well-crystallized varieties of the igneous and metamorphic rocks, the "clay chlorites" appear to contain more iron, are finer-grained, and have poorly organized structures with unbalanced charges.

e. *The Vermiculite Group.* The structure and properties of the vermiculite minerals appear intermediate between those of the chlorites and montmorillonites. The untreated minerals have a unit thickness of 14–15Å and are composed of talclike layers between which are located sheets of hydrated magnesium ions. Unlike the chlorites, however, when the vermiculites are heated, they readily lose water from the interlayered sheets and the structure contracts to 9–10Å, while the essential difference between montmorillonite and vermiculite is the failure of the latter to expand beyond 15Å on water or glycol saturation. The positive charge deficiency in the talclike layers of vermiculite is intermediate between those of chlorite and montmorillonite and herein lies the fundamental relationship of the three mineral groups.

"Clay vermiculites" have been recorded in sediments and soils but, like the clay chlorites, little is known of their true structures and properties.

f. *Mixed-Layered Clay Minerals.* In the previous sections attention was drawn to the similarities in structure between the illites, montmorillonites, chlorites, and vermiculites. Although these minerals differ from one another in the extent of the positive charge deficiency of the talc- or pyrophyllitelike layers and in the nature of the interlayered cations, nevertheless, the similarities are such that interstratifications of one mineral with another are possible, indeed common. The interstratifications may be of two types: (i) a regular stacking of the layers resulting in a repeating pattern, for example, A B A B A B A B, or (ii) an irregular stacking of the layers, so that the pattern is not repeating, for example, A B A A B A B B A, where A and B are any two members of the four mineral groups. Illite-montmorillonite and chlorite-vermiculite mixed layers are particularly common. The properties of the mixed crystals are similar to mixtures of the discrete minerals.

g. *The Palygorskite (Attapulgite)–Sepiolite Group.* Minerals of the paly-gorskite–sepiolite group are unique among the clay minerals in that the structures are not based on the phyllosilicates but rather consist of modified amphibole double chains. Magnesium is the principal octahedral cation although replacement by aluminum, ferrous, and ferric ions is common.

Apart from a few isolated hydrothermal occurrences, minerals of the group are virtually restricted to playa lake environments and arid soils (Rogers *et al.*, 1956; Loughnan, 1960; Vanden Heuval, 1964), and are unknown in older sediments.

2. *The Residual Oxides and Hydroxides*

Of the nine common cations in silicate rocks only four, Si^{4+}, Al^{3+}, Fe^{3+}, and Ti^{4+}, form stable oxides or hydroxides or both in the weathering environment.

a. *Oxides of Silicon.* Silica exists in three crystalline polymorphic forms, *quartz*, *tridymite*, and *cristobalite*, and each has a low- and a high-tempera-ture modification termed respectively α and β. Natural occurrences of tridy-mite and cristobalite are essentially restricted to high-grade contact meta-morphic rocks (sanidinite facies), although cristobalite has been recorded as a constituent of clays derived from the alteration of volcanic ash known as bentonite (Gruner, 1940). Alpha quartz is the characteristic crystalline form of silica in sedimentary rocks and soils.

Chalcedony is the term applied to a number of varieties of cryptocrystalline silica, ranging from spherulitic radiating masses to fine aggregates. Chalce-dony gives the X-ray pattern of quartz but differs from the latter mineral in some of its optical properties (White and Corwin, 1961). It is common as vesicle and geode infillings in volcanic rocks and as the crystalline product of silica deposited in cavities in sedimentary rocks and in the B horizon of certain soils.

In addition to these crystalline forms of silica, amorphous (i.e., amorphous to X-radiation) hydrated varieties of silica are known and include materials such as *opal*, *diatomite*, and so on. Because of the failure of these materials to record an X-ray pattern, they are difficult to recognize and it is possible their occurrence in sediments and soils is more widespread than generally believed.

b. *Hydroxides of Aluminum.* Aluminum occurs naturally as the oxide *corundum*, as the trihydrate *gibbsite*, and as the monohydrates *boehmite* and *diaspore*. Corundum (Al_2O_3) is characteristic of undersaturated syenites, ultrabasic rocks, and contact-metamorphosed limestones, and it is not norm-ally encountered in weathering environments. Diaspore $AlO(OH)$ has been reported in sedimentary fireclays from Pennsylvania (Bolger and Weitz, 1952) and Missouri (Keller, 1952) but has not been observed as a weathered pro-duct in soils. The pressure-temperature equilibrium curves of Roy and Osborn (1952) for the system Al_2O_3–SiO_2–H_2O indicate that diaspore should form only at elevated temperatures. Gibbsite, $Al_2(OH)_6$, and boehmite, $AlO(OH)$, on the other hand, form at ordinary pressures and temperatures and both are relatively common in highly leached soils of tropical areas.

The structure of gibbsite is similar to the octahedral sheets in the clay minerals and consists of close-packed octahedra of hydroxyl ions bonded by aluminum in sixfold coordination. Only two thirds of the possible cation sites are occupied. Boehmite, on the other hand, has a more complex structure comprised of a double sheet of octahedra with aluminum ions at their centers. The oxygens in boehmite are of two types: those in the center of the structure are shared by four octahedra whereas those on the outside are linked to only two. The sheets are held together by hydrogen bonding. The difference in structures between boehmite and gibbsite results in a contrast in behavior of the two minerals. Boehmite is remarkably resistant to attack by alkalies and acids whereas gibbsite is readily soluble in solutions at pH values above 10 and below 4.

c. *Oxides and Hydroxides of Iron*. Unlike aluminum, iron does not occur as the trihydrate, but rather forms two monohydrates and several oxide minerals which, with the exception of magnetite, are common residual secondary products derived from the chemical breakdown of rocks. *Goethite*, the more common of the two monohydrates, is isostructural with diaspore, and *lepidocrocite*, the other monohydrate, is analogous to boehmite. The oxide minerals include *magnetite*, a member of the spinel group with the composition $FeO \cdot Fe_2O_3$, and the two polymorphic forms of ferric oxide, *hematite* and *maghemite*. Magnetite is characteristic of igneous and metamorphic rocks and, although it may persist as a detrital grain in sediments, generally it is unstable in the weathering environment and is readily converted to the higher oxidized states of maghemite and hematite. Maghemite also has a spinellike structure and like magnetite displays ferromagnetic properties. It occurs in laterites, where generally it is intimately mixed with hematite. Rooksby (1961) has considered maghemite to be metastable with respect to hematite and to convert rapidly to the latter mineral at temperatures in excess of 400°C.

In most soils and sedimentary rocks much of the iron oxide present contains indefinite amounts of water and appears amorphous to X-radiation. The term *limonite* has been applied to this poorly organized material.

The iron oxide and hydroxide minerals are particularly sensitive to changes in the redox potential of the environment. In the presence of organic matter or below the water table, ferric minerals tend to be destroyed whereas, in the oxidized parts of the environment, ferrous compounds are particularly unstable.

d. *Oxides of Titanium*. Compared with the preceding, the residual oxides of titanium are not particularly abundant. Nevertheless, most parent materials contain small amounts of titanium-bearing minerals and the processes of leaching tend to concentrate the element in the soil. In tropical regions of high rainfall, titania may become a major constituent of the surface horizon and exceed 25% of the total mineral content (Sherman 1952).

Of the three polymorphic forms of TiO_2, the tetragonal minerals, *rutile* and *anatase*, are common as residual products of weathering whereas *brookite*, the orthorhombic equivalent, is rare. Much of the rutile in soils and sediments is of primary origin. Anatase, on the other hand, is generally formed from titania released from primary minerals such as sphene, titaniferous magnetite, and pyroxenes, by the removal of more mobile constituents of these minerals. Anatase occurs either as pseudooctahedral crystals or as polycrystalline aggregates referred to as *leucoxene*. *Ilmenite*, the oxide of iron and titanium, is moderately stable in the weathering environment and may persist as detrital grains in the derived sediments.

REFERENCES

Bates, T. F. (1952), Interrelationships of structure and genesis in the kaolinite group. Problems of clay and laterite genesis, *Am. Inst. Min. Met.*, pp. 144–153.

Bolger, R. C. and J. H. Weitz (1952), Mineralogy and origin of the Mercer Fireclay of North Central Pennsylvania. Problems of clay and laterite genesis. *Am. Inst. Min. Met.*, pp. 81–93.

Bragg, W. L. (1937), *Atomic Structure of Minerals*. Cornell Univ. Press, Ithaca, New York.

Brindley, G. W. (1951), X-ray identification and crystal structure of the clay minerals. Min. Soc. Great Britain, London.

Burst, J. F. (1958), Glauconite pellets, their mineral nature and application to stratigraphic interpretations. *Bull. Am. Assoc. Petrol. Geologista* **42**, 315–327.

Cartledge, G. H. (1928), Studies in the periodic system. *J. Chem. Soc. Am.*, pp. 2855–2872.

Craig, D. C. and F. C. Loughnan (1964), Chemical and mineralogical transformations accompanying the weathering of basic volcanic rocks from New South Wales, *Australian J. Soil. Res.* **2**, 218–234.

Grim, R. E. (1953), *Clay Mineralogy*. McGraw-Hill, New York.

Grim, R. E. (1962), *Applied Clay Mineralogy*. McGraw-Hill, New York.

Grim, R. E., R. M. Bray and W. F. Bradley (1937), The mica in argillaceous sediments, *Am. Mineralogist* **22**, 813–829.

Gruner, J. W. (1940), Abundance and significance of cristobalite in bentonite and fullers earth, *Econ. Geol.* **35**, 867–875.

Hendricks, S. B. (1942), Lattice structure of clay minerals, *J. Geol.* **50**, 276–290.

Hendricks, S. B. and C. S. Ross (1941), Chemical composition and genesis of glauconite and celadonite. *Am. Mineralogist* **26**, 683–708.

Hendricks, S. B. and M. E. Jefferson (1938), Kaolin and talc-pyrophyllite hydrates; water sorption of clays. *Am. Mineralogist* **23**, 863–875.

Hofmann, U., K. Endell and D. Wilm (1933), Kristallstruktur und Quellung von Montmorillonit, *Z. Krist.* **86**, 340–348.

Keller, W. D. (1952), Observations on the origin of Missouri high alumina clays. Problems of clay and laterite genesis. *Am. Inst. Min. Met.*, pp.115–134.

Krauskopf, K. B. (1959), The geochemistry of silica in sedimentary environments. Silica in sediments symposium, *Soc. Econ. Paleontologists Mineralogista, Spec. Publ.* **7**, 4–19.

Levinson, A. A. (1955), Polymorphism between illites and hydrous micas, *Am. Mineralogist* **40**, 41–49.

Loughnan, F. C. (1960), Further remarks on the occurrence of palygorskite at Redbank Plains, Queensland, *J. Roy. Soc. Queensland* **71**, 43–50.

Loughnan, F. C. and P. Bayliss (1961), The mineralogy of the bauxite deposits near Weipa, Queensland, *Am. Mineralogist* **46**, 209–217.

MacEwan, R. M. C. (1961) The X-ray identification and crystal structures of clay minerals, Min. Soc. Great Britain, London, pp. 143–207.

Maegdefrau, E., and U. Hofmann (1937), Die Kristallstruktur des Montmorillonits, *Z. Krist* **98**, 299–323.

Marshall, C. E. (1935), Layer lattices and base exchange clays, *Z. Krist* **91**, 433–449.

Marshall, C. E. (1964), *The Physical Chemistry and Mineralogy of Soils*. Vol. 1. *Soil Materials*. Wiley, New York.

McMurchy, R. C. (1934), Crystal structure of chlorites, *Z. Krist.* **88**, 420–432.

Pauling, L. (1940), *The Nature of the Chemical Bond*, 3rd ed. Cornell Univ. Press, Ithaca, New York

Radoslovich, E. W. (1960), The structure of muscovite, *Acta. Cryst.* **13**, 919–932.

Ramberg, H. (1954), Relative stabilities of some simple silicates as related to the polarization of the oxygen ions. *Am. Mineralogist* **39**, 256–271.

Rogers, L. E. R., J. P. Quirk and K. Norrish (1956), Occurrence of an aluminum sepiolite in a soil having unusual water relationships, *J. Soil Sci.* **7**, 177–184.

Rooksby, H P. (1961), The X-ray identification and crystal structures of the clay minerals, Min. Soc. Great Britain, London, pp. 354–392.

Ross, C. S., and S. B. Hendricks (1945), Minerals of the montmorillonite group, *U.S. Geol. Surv. Profess. Paper* **205B**, 23–77.

Ross, C. S., and E. V. Shannon (1926), Minerals of bentonite and related clays and their physical properties, *J. Am. Ceram. Soc.* **9**, 77–96.

Roy, R., and E. F. Osborn (1952), Studies in the system alumina-silica-water. Problems of clay and laterite genesis, *Am. Inst. Min. Met.*, pp. 76–80.

Sherman, G. D. (1952), The titanium content of Hawaiian soils and its significance, *Proc. Soil Soc. Am.* **16**, 15–18.

Vanden Heuval, R. C. (1964), The occurrence of sepiolite and attapulgite in the calcareous zone of a soil near Las Cruces, New Mexico, *Proc. Nat. Conf. Clays and Clay Minerals* **13**, 193–208.

Weaver, C. E. (1958), Geologic interpretation of argillaceous sediments, *Bull. Am. Assoc. Petrol. Geologists* **42**, 254–271.

White, J. F., and J. F. Corwin (1961), Synthesis and origin of chalcedony, *Am. Mineralogist* **46**, 112–119.

III. The Chemistry of Weathering

INTRODUCTION

Chemical weathering results from a change in chemical environment. Minerals that have formed under magmatic, hydrothermal, metamorphic, or sedimentary conditions are rendered potentially unstable when exposed to the atmosphere. They are vulnerable to attack by water, oxygen, and carbon dioxide, and the reactions, which are exothermic (i.e., liberate heat), tend to proceed spontaneously. Water penetrates through pores, cleavages, and other micro openings in the minerals and dissolves the more soluble constituents. As these processes intensify, the residue becomes progressively enriched in the less soluble constituents as well as in oxygen and hydroxyl groups. Ultimately, crystallization of the residue results in the development of new mineral phases which are in more stable equilibrium with the prevailing atmospheric conditions.

Since the reactions involved are subject to the laws of chemical equilibria, the breakdown of a mineral can proceed beyond the state of equilibrium only if components are added and (or) removed from the system. Thus in the alteration of potash feldspar to kaolinite according to the following reaction

$$\underset{\text{potash feldspar}}{2KAlSi_3O_8} + \underset{\text{water}}{3H_2O} \rightarrow \underset{\text{kaolinite}}{Al_2Si_2O_5(OH)_4} + \underset{\text{silica}}{4SiO_2} + \underset{\text{potash}}{2KOH}$$

all the potash must be lost in solution. If some is retained, illite and not kaolinite will be the residual product:

$$\underset{\text{potash feldspar}}{3KAlSi_3O_8} + \underset{\text{water}}{2H_2O} \rightarrow \underset{\text{illite}}{KAl_2(Al, Si_3)O_{10}(OH)_2} + \underset{\text{silica}}{6SiO_2} + \underset{\text{potash}}{2KOH}$$

It is apparent, therefore, that a mineral can develop as a result of weathering only if the atoms and ions essential to the formation of the mineral are present within the weathering environment or are introduced, and remain or are rendered immobile.

Immobility may be due to the low solubility of the atoms and ions in the prevailing environmental conditions—this is generally true in the case of aluminum, ferric iron, and titanium—or it may result from incomplete breakdown of the mineral structure. Many silicate minerals subjected to chemical weathering are reduced to smaller units which still retain, in part, the original polymerized structure along with some of the potentially mobile cations entrapped therein. The micas afford an excellent example in this respect. Although much of the potassium may be removed from between the layers of these minerals, the layers themselves, composed of polymerized silica tetrahedra and potentially mobile cations, are often remarkably per-

27

sistent. Moreover, even where complete collapse of the structure does eventuate, the rate of release of potentially mobile constituents may exceed their rate of solution so that at least some of these mobile constituents may become incorporated in the crystalline phases which develop from the residue.

In brief, it would appear that three simultaneous processes are involved in the weathering of the silicate minerals.

(a) The breakdown of the parent mineral structures with the concomitant release of cations and silica. The "released" silica may be reduced to the monomeric form, that is, a molecularly dispersed state, or it may persist in a polymerized form, that is, much of the original structure is retained.

(b) The removal in solution of some of the "released" constituents.

(c) The reconstitution of the residue with components from the atmosphere such as water, oxygen, and carbon dioxide, to form new minerals which are in stable or metastable equilibrium with the environment.

This is undoubtedly the concept Merrill (1906) had in mind when he wrote, "In glancing over the analyses, it is at once apparent that hydration is an important factor, the amounts of water increasing as the decomposition advanced. In the earlier stages of degeneration it is doubtless the most important factor. There is, moreover, among the siliceous crystalline rocks in every case, a loss of lime, magnesia and the alkalies and a proportional increase in the amounts of alumina and sometimes ferric oxide, though the apparent gain may be in some cases due to the change in condition from ferrous to ferric oxide. As a whole, however, there is a very decided loss of materials. Among siliceous rocks this loss, so far as is shown by available analyses and calculations, rarely amounts to more than 60% of the entire mass."

MECHANISM OF BREAKDOWN OF PARENT MINERAL STRUCTURES

Jenny (1950) has given perhaps the clearest exposition of the mechanism of breakdown of the parent mineral structures. He has pointed out that, although the sum of the negative charges within any crystal must equate the sum of the positive charges in accordance with Pauling's rules, exposed atoms and ions at the surfaces possess unsaturated valencies and, when these are brought into contact with water, hydration occurs through the attraction of the water dipoles to the charged surfaces. The attractive forces may polarize the water dipoles to such an extent that dissociation into hydrogen and hydroxyl ions ensues, thus permitting the hydroxyl ions to bond to exposed cations and the hydrogen ions to oxygens and other negative ions. Simultaneously, the hydrogen ions may replace cations at the mineral surface and the release of the latter causes the pH of the liquid phase to rise. Using feldspar as an example, Jenny proposed the following reaction:

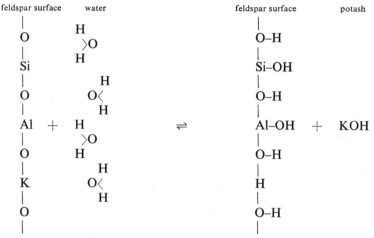

The conversion of the oxygens to hydroxyl groups and the removal of potassium permits aluminum, which originally is present in tetrahedral coordination with oxygen, to assume its preferred octahedral coordination and, as a result of these processes, the surface layers are rendered unstable and various polyhedra are released to the surrounding water. Initially the released polyhedra may form amorphous colloids but with aging they become oriented into the structures of the secondary minerals such as clays and oxides.

There is much experimental data to support Jenny's hypothesis. Tamm (1930) ground various feldspars in both distilled water and carbonic acid and determined the pH values of the solution. He found that decomposition of the feldspars, as determined by the increase in pH values and the release of silica, alumina, and the alkalies to the solutions, depended on the physical action of producing fresh surfaces. However, a limit was reached beyond which continued grinding failed to produce further loss of constituents and from this point onward the pH remained essentially constant.

Stevens and Carron (1948) carried out similar experiments. However, they examined a wider array of minerals and found those containing the alkalies and alkaline earths invariably yielded abrasion pH values on the alkaline side of neutrality whereas those devoid of these ions produced acid values (Table 4). Apparently the processes of hydration, hydrolysis, and hydrogen replacement of cations continue until a state of equilibrium is attained. At this point just as many ions reenter the structure as leave it according to the following equation:

$$M^+[\text{mineral}]^- + H^+ OH^- \rightleftharpoons H^+[\text{mineral}]^- + M^+ OH^-$$

Garrels and Howard (1959) determined the pH values of suspensions of muscovite and adularia, both previously ground to pass a 200-mesh screen. They interpreted their results as indicating that the initial reaction of minerals with water produces a layer of alteration which is structurally disrupted at

TABLE 4

Abrasion pH Values for Some Minerals[a]

Mineral	Formula	Abrasion pH
Silicates		
Actinolite	$Ca_2(MgFe)_5Si_8O_{22}(OH)_2$	11
Diopside	$CaMg(SiO_3)_2$	10, 11
Olivine	$(MgFe)_2SiO_4$	10, 11
Tremolite	$Ca_2Mg_5Si_8O_{22}(OH)_2$	10, 11
Augite	$Ca(MgFeAl)(AlSi)_2O_6$	10
Hornblende	$(CaNa)_2(MgFeAl)_5(AlSi)_8O_{22}(OH)_2$	10
Leucite	$KAlSi_2O_6$	10
Albite	$NaAlSi_3O_8$	9, 10
Aegerine	$(NaCa)(FeAlMg)Si_2O_6$	9
Oligoclase[b]	$Ab_{90-70}An_{10-30}$	9
Talc	$Mg_3Si_4O_{10}(OH)_2$	9
Anthophyllite	$(MgFe)_7Si_8O_{22}(OH)_2$	8, 9
Biotite	$K(MgFe)_3(AlSi_3)O_{10}(OH)_2$	8, 9
Labradorite[b]	$Ab_{50-30}An_{50-70}$	8, 9
Microcline	$KALSi_3O_8$	8, 9
Anorthite	$CaAl_2Si_2O_8$	8
Hypersthene	$(MgFe)_2Si_2O_6$	8
Muscovite	$KAl_2(AlSi_3)O_{10}(OH)_2$	7, 8
Orthoclase	$KAlSi_3O_8$	8
Andalusite	Al_2SiO_5	7
Montmorillonite	$Al_2Si_4O_{10}(OH)_2 \cdot nH_2O$	6, 7
Halloysite	$Al_2Si_2O_5(OH)_4$	6
Pyrophyllite	$Al_2Si_4O_{10}(OH)_2$	6
Kaolinite	$Al_2Si_2O_5(OH)_4$	5, 7
Oxides		
Boehmite	$AlO(OH)$	6, 7
Gibbsite	$Al(OH)_3$	6, 7
Quartz	SiO_2	6, 7
Hematite	Fe_2O_3	6
Carbonates		
Magnesite	$MgCO_3$	10, 11
Dolomite	$CaMg(CO_3)_2$	9, 10
Aragonite	$CaCO_3$	8
Calcite	$CaCO_3$	8
Siderite	$FeCO_3$	5, 7

[a] After Stevens and Carron (1948).
[b] Ab=albite; An=anorthite.

the outer margin but grades inward to a zone where the original structure is preserved but the potassium ions are replaced by hydrogen. Moreover, their experiments indicated that "at 25°C an H-feldspar or H-mica structure is favored over a K-feldspar or K-mica structure except in solutions in which the ratio of a_{K^+}/a_{H^+} exceeds 10^{9-10} or 10^{7-8} respectively."

Frederickson (1951) has envisaged the mechanism of weathering of albite feldspar "as a process whereby the hydrogen ions of crystalline water are base exchanged for the Na^+ of albite. The water layers on the feldspar exist as a crystalline network with a high degree of order. The small hydrogen ions from the water enter and upset the neutrality of the crystal. The crystal attempts to become neutral again by rejecting the Na^+ which is less strongly held than the invading hydrogen. The ion substitution causes the crystal to expand and the chemical activity is increased which hastens the eventual collapse of the crystal. The residual products of decomposition are gels or insoluble silica depending on the Al/Si ratio existing in the system being considered."

McConnell (1951) has raised a number of objections to Frederickson's proposals while the concept of similarity in structure between crystalline water and feldspar cannot be extended to include all silicate groups.

According to Fieldes and Swindale (1954), with the exception of the micas, all primary silicates must pass through an amorphous stage in the transition to secondary crystalline phases. If the term "amorphous" implies "a state devoid of all organized structure," then it becomes difficult to understand how illite, for example, could develop from potash feldspar. One would expect the reduction of the feldspar to individual tetrahedra to result in the complete solution of potassium.

De Vore (1959) has discussed the mechanism of formation of layered lattice minerals, such as the micas, from feldspar. He has pointed out that decomposition of the feldspar structure releases chains which have a certain degree of stability and retain the original Si–Al ordering of the tetrahedra. If these released chains are from the (100) and (010) surfaces of the feldspar they can polymerize directly into tetrahedral sheets of composition $(AlSi_3)O_{10}$. Combination with octahedral cations (e.g., Al^{3+}, Mg^{++}, Fe^{++}, and Fe^{3+}) and K^+ could lead to the formation of clay minerals. However, if the released chains break down to individual tetrahedra, aluminum contained therein would be expected to assume its preferred octahedral coordination. In which case, layered lattice silicates requiring at least part of the aluminum in the tetrahedral sheets could not develop or, more specifically, formation of illite, montmorillonite, and chlorite from feldspar is possible only if the secondary products inherit some of their structure from the parent mineral. Kaolins are to be expected where the parent minerals break down to individual tetrahedra.

From the above discussion two conclusions clearly emerge.

(a) Hydrogen ions produced by the hydrolyzing action of mineral surfaces play an important role in the breakdown of silicate structures. The size of these ions permits easy penetration into crystal lattices and, once they have gained admission, the high charge-to-radius ratio, which is greater than for any other ion, has a marked disrupting effect on the charge balance within the lattice.

(b) The persistence of aluminum in tetrahedral coordination and the retention of highly mobile cations in the secondary mineral products indicate that much of the parent mineral structure is inherited by the secondary products.

FACTORS INFLUENCING THE MOBILITIES OF THE COMMON METALLIC IONS

In the previous section the mechanism of chemical weathering of silicate minerals was discussed and it was concluded that breakdown proceeds through solution or partial solution of some of the constituent cations of the mineral. However, not all cations are taken into solution with the same ease. Some tend to be readily lost to the percolating ground waters whereas others are resistant and become progressively concentrated in the residue. Moreover, variations in the physicochemical factors may result in solution of a particular cation in certain parts of a weathering environment and pre-cipitation elsewhere.

In the following, consideration is given to the nature of the physicochemical factors and to their influence on the solubilities of the metallic ions with the primary objective of determining the independent variables which control the loss or retention of these various ions. For simplicity, discussion is restricted to the more common cations occurring in silicate minerals, namely, Si^{4+}, Al^{3+}, Fe^{3+}, Fe^{++}, Ti^{4+}, Mg^{++}, Ca^{++}, Na^+, and K^+.

Influence of pH on the Mobilities of the Common Cations

According to Mason (1952), the pH values of natural waters normally lie between 4 and 9. It is necessary, therefore, to consider the influence of varia-

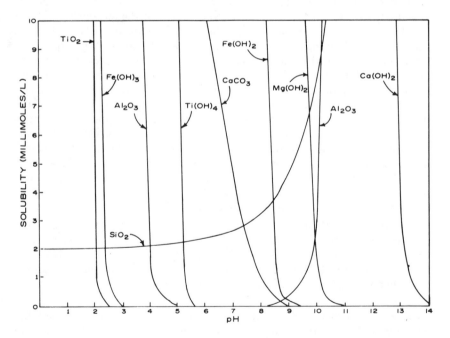

Figure 15.   Solubility in relation to pH for some components released by chemical weathering.

tions within this range on the mobilities of the principal components involved in weathering of the silicate minerals. In Fig.15, which represents a compilation from many sources, the solubilities of the more important components are plotted against pH. The curves for Al_2O_3, SiO_2, TiO_2, and $CaCO_3$ have been taken directly from Correns (1949), Krauskopf (1958), Vinogradov (see Bardossy, 1959), and Correns (1949), respectively, while the remainder have been calculated from their solubility products at 20–25°C (*Chem. Soc. London Spec. Publ.* **17** (1964)).

This diagram shows that within the "normal" pH range of ground waters, $Ca(OH)_2$, $Mg(OH)_2$, and the alkalies (not shown) are completely soluble whereas TiO_2, $Fe(OH)_3$, and Al_2O_3 remain insoluble and hence cannot be mobilized. The solubility of silica, although low, is constant and unaffected by variations in the pH. The only components, therefore, whose mobility is influenced by changes in the environmental pH, that is, the only ones that are pH-dependent, are $Ti(OH)_4$, $CaCO_3$, and $Fe(OH)_2$. It should be noted that the curve for SiO_2 refers to amorphous silica or silica released from silicates and not quartz, which, according to Krauskopf (1958), has a solubility of approximately one-tenth that of amorphous silica.

However, exceptional environments in which the pH of the ground waters lies outside the "normal" range are also known. Thus, for example, in certain playa lakes, where the rocks subject to alteration are rich in minerals with high abrasion pH values, such as olivine, augite, and nepheline, the waters may become particularly alkaline, sufficiently to cause solution of alumina as aluminates and precipitation of magnesia. Such an environment would tend to favor the formation of minerals of the palygorskite-sepiolite group (Loughnan, 1960), generally in association with dolomite. This is probably also the environment of formation of authigenic zeolites (Keller, 1957; Hay and Moiola, 1963; Loughnan, 1966).

Again, horizons containing abundant organic matter in well-drained soils, such as the podsols, may attain pH values below 4 (see Chapter VI). Under these conditions, alumina is again rendered potentially mobile and may migrate down the soil profile to less acid regions where precipitation and enrichment take place. This is well illustrated in Fig. 16, in which the silica-to-alumina ratios for the clay colloids from a podsol soil are plotted against depth. A similar plot was also made for a lateritic soil and it is interesting to compare the two curves. Despite the fact that both soils are well leached and contain similar clay minerals, a marked contrast exists, apparently brought about by the very low pH values attained by the organic-rich layers in the podsol.

Several attempts have been made to determine empirically the order of loss of constituents. Thus Polynov (1937) compared "the average composition of mineral matter dissolved in rivers whose catchment basins are mostly in areas of massive rocks" with "the average composition of igneous rocks" (Table 5) and obtained an order of loss corresponding to:

$$Ca^{++} > Na^+ > Mg^{++} > K^+ > SiO_2 > Fe_2O_3 > Al_2O_3$$

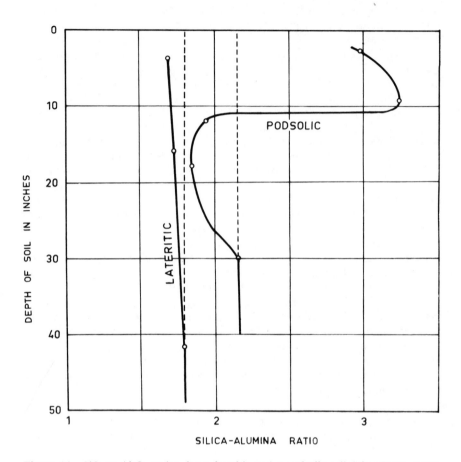

Figure 16.   SiO₂-to-Al₂O₃ ratios for a lateritic and a podsolic soil (after Jenny, 1950).

The order is in very good agreement with that obtained by Goldich (1938) from the study of the weathering of a granite gneiss in Minnesota (p. 97) and by Tiller (1958) from the weathering of basaltic materials in South Australia. Perhaps the most interesting feature of this series is the apparent anomalous position of potassium. This point is discussed later in a section dealing with fixation of cations (p. 45).

The Hydrogen Ion Concentration as an Independent Variable

The role played by the hydrogen ion in weathering cannot be overstressed. Not only does it initiate disruption of crystal structures but also, as discussed above, its concentration influences the solution and precipitation of some of

TABLE 5
Relative Mobilities of Rock Constituents[a]

Constituent	Average composition of igneous rocks (a)	Average composition of the mineral residue of river waters (b)	Relative mobility of elements and compounds[b] (c)
SiO_2	59.09	12.80	0.20
Al_2O_3	15.35	0.90	0.02
Fe_2O_3	7.29	0.40	0.04
Ca	3.60	14.70	3.00
Mg	2.11	4.90	1.30
Na	2.97	9.50	2.40
K	2.57	4.40	1.25
Cl	0.05	6.75	100.00
SO_4	0.15	11.60	57.00
CO_3	—	36.50	—

[a] After Polynov (1937); used by permission of George Allen and Unwin Ltd.
[b] The values c represent the ratio b/a adjusted to chlorine = 100.

the released ions. However, the objective here is to determine whether the hydrogen ion concentration is an independent variable in weathering or whether it is the effect of other factors, rather than to assess its function as an agent of chemical weathering.

Alkaline environments are created by the presence of alkali and alkaline earth ions which may have been introduced from extraneous sources but, more commonly, have been released from the minerals undergoing weathering. In areas close to the coast, deposition of salt is generally appreciable. Walker (1960) has shown that along the sea front in the vicinity of Sydney, Australia more than 300 lb/acre may be deposited in a 2-month period. However, the rate of deposition decreases rapidly inland and 10 miles from the coast only 9·3 lb/acre were deposited during the same period. Nevertheless, even in remote areas, if the climate is arid or semiarid and leaching ineffectual, salt accumulations may result. Many of the soils of southern and western Australia are contaminated with sodium chloride undoubtedly of this origin. Again, in low-lying areas, surface wash and subsoil drainage from surrounding hills may introduce alkali and alkaline earth metals into the weathering environment. This point is exemplified by a comparison of peats formed in moorland and fen areas. According to Robinson (1949), the waters of the upland peats are recharged mainly by direct rainfall and, as a result, dissolved salts are sparse and the pH is distinctly acid. The low-lying fen peats, on the other hand, receive drainage from the surrounding areas and the waters are generally charged with salts which render the swamp waters neutral or even alkaline.

However, the major source of the alkaline ions in the weathering environment is undoubtedly the parent minerals themselves. The work of Stevens and Carron (1948) and Keller et al. (1963) indicates that the weathering surfaces of minerals containing alkali or alkaline earth ions have very high pH values (Table 4). Whether the environment as a whole will have similar high pH values depends on how effectively the released alkalies and alkaline earth are removed from the system. As the degree of leaching increases these ions tend to be flushed out and the environmental pH, as distinct from the mineral-surface pH, is lowered toward neutrality. Where the rate of removal of these ions exceeds their rate of release from the decomposing minerals, the system assumes an acid pH value.

Whereas water is the ultimate source of hydrogen ions in the weathering environment, the processes by which these ions are concentrated are often not so obvious.

A high concentration of hydrogen ions may result from the oxidation of sulfide minerals such as pyrite or marcasite, in the presence of water,

$$\underset{\text{pyrite}}{2FeS_2} + \underset{\text{water}}{2H_2O} + \underset{\text{oxygen}}{7O_2} \rightleftharpoons \underset{\text{ferrous sulfate}}{2FeSO_4} + \underset{\text{sulfuric acid}}{2H_2SO_4}$$

and frequently in areas of sulfide mineralization the ground waters attain very low pH values. Sulfide minerals, however, are not abundant in silicate rocks and their influence on rock weathering, although locally often quite marked, in general is of little consequence.

Hydrogen ions can be furnished also by reaction of carbon dioxide with water to yield carbonic acid. The normal atmosphere contains 0.03% by volume of carbon dioxide but, according to Keller (1957), soil atmospheres may contain more than 10 times this concentration. Carbonic acid, however, is weakly ionized and, under the normal partial pressure of the atmosphere, a saturated solution assumes a pH value a little below 6. Consequently, the presence of carbon dioxide does not have a particularly marked effect on the environmental pH. It plays a far more important role in reacting with carbonates to form soluble bicarbonates; for example, $CaCO_3 + CO_2 + H_2O \rightleftharpoons Ca(HCO_3)_2$.

According to Williams and Coleman (1950), plant roots, particularly those low in the order of evolution, generally have a pH value below 4, and some possibly as low as 2. Keller and Frederickson (1952) believe that hydrogen ions exist in a diffuse double layer about the roots and that weathering of silicate minerals is accomplished through exchange reactions in which metallic ions of the minerals replace the hydrogen ions and are ultimately assimilated by the plants. At the same time, the hydrogen ions become concentrated at the mineral surfaces. However, the rate of supply of hydrogen ions is restricted to the rate of assimilation of metallic ions and, although there can be little doubt as to the importance of this mechanism, it is difficult to attribute the entire concentration of hydrogen ions, at least, in some soils, solely to this means.

Because of unsaturated valencies of atoms and ions at their surfaces, isomorphous substitution within their lattices, or exposed hydroxyl groups, clay minerals possess negative charges. They endeavor to attain neutrality by adsorption of cations, which, although bonded to the surfaces of the minerals, may be readily exchanged for other cations in solution. The measure of the ability of clays to adsorb cations is termed the *cation exchange capacity* (cec) and is expressed numerically as the number of milliequivalents (meq) adsorbed per 100 g of clay. Where the adsorbed ion is hydrogen, the clays behave as weak acids or "colloidal acids" and a close relationship exists between the cation exchange capacity of the clay mineral and the pH it assumes upon saturation with hydrogen ions. Thus, kaolinite with a cec of 5–10 meq/100 g attains pH values of the order of 4–5 on saturation with hydrogen ions, whereas H^+-montmorillonite with a cec of 80–100 meq/100 g may have values as low as 3.0 (Fig. 17). Colloidal organic material also possesses charged surfaces capable of adsorbing hydrogen ions and yielding weak "colloidal acids" similar to those formed by the clay minerals.

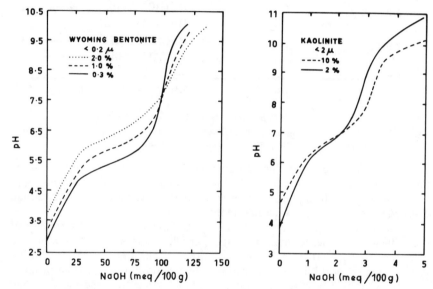

Figure 17. Titration curves for H^+-bentonite and H^+-kaolinite (after Marshall and Krinbill, 1942; used by permission of The Williams and Wilkins Co.).

The effectiveness of these "colloidal acids" in attacking silicate structures was demonstrated by Graham (1941a, 1941b). He mixed humic acid, H^+-agar, H^+-bentonite, and acetic acid with primary minerals such as anorthite and in each case found that the pH increased with time. Undoubtedly the increase was due to the release of alkalies and alkaline earths from the parent structures. H^+-Bentonite proved more effective than acetic acid in releasing calcium from anorthite. However, reference to Graham's diagram (Fig. 18) shows that the curves for pH against time are still rising after a period of 110 days and hence equilibrium for the reactions was not attained. Since the calcium released from the anorthite forms a strong base whereas the acids used are weak, the equilibrium pH values should be on the alkaline side and prob-

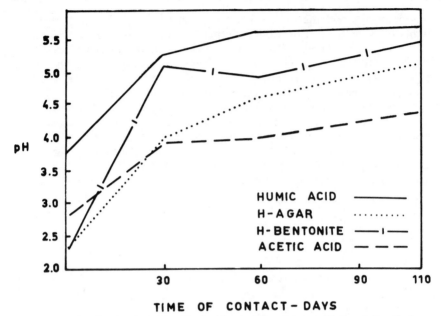

Figure 18.   Showing the change in pH of acid-anorthite mixtures with time (after Graham, 1941b; used by permission of The Williams and Wilkins Co.)

ably would approach the value of 8 given by Stevens and Carron (Table 4) as the· abrasion pH for anorthite. This would suggest that H+-bentonite and humic acid do not bring about any greater total destruction of anorthite than distilled water; they merely accelerate the process.

Graham's techniques were employed by McClelland (1950) in a study of the influence of time, temperature, and particle size distribution on the rate of release of bases from a number of primary silicate minerals including olivine, augite, hornblende, and various feldspars and micas. The primary minerals were first fractionated into a series of particle size gradings and then treated with H+-bentonite. Quartz was added to facilitate filtering. Similar experiments were carried out using Ca++-bentonite as the weathering agent. From the results McClelland concluded:

"1. The hydrolysis of ground minerals is appreciable, but the extent of hydrolysis of bases varies for the different minerals and for the different ions in the same mineral.

"2. The rate of release of bases from minerals increases with decreasing particle size, but the extent to which particle size influences base release varies with different minerals.

"3. Increasing the ratio of Ca:H on the colloidal complex decreases the rate of base release.

"4. The rate of base release increases with temperature, but the effect of increased temperature decreases with time.

"5. Fresh minerals release bases at a fast rate, but this rate rapidly decreases with time. It appears that residual primary weathering products retard the release of bases from minerals, presumably by accumulating close to the weathering surface.

"6. Apparently the release of bases from olivine, augite, hornblende, albite, labradorite, microcline, anorthoclase, and phlogopite is accompanied by the breakdown of the crystal lattice of these minerals. The release of K from biotite appears to proceed faster than decomposition of that mineral would indicate.

"7. The rate of the release of two ions present in the same mineral tends to approximate the ratio in which these ions are present in the unweathered mineral with increasing time.

"8. The silt and coarse clay fractions of the K-bearing minerals studied release K at appreciable rates. All of these minerals could be important sources of K if present in soils.

"9. The order with which the minerals studied release bases approximates Goldich's stability series [see p. 53] with the exception of muscovite."

Up to this point, discussion of the cause of variations in the pH values of weathering environments has been confined to closed systems, that is, there has been no removal of components. Under these conditions, as in many arid or semiarid areas, the strong bases such as sodium, potassium, calcium, and magnesium are retained and the environment is alkaline. However, if the system is open and these products are removed at a rate equal to or greater than their release from the parent mineral, then the residue becomes progressively more acid. In the alteration of anorthite according to the following reaction,

anorthite   montmorillonite

$$3CaAl_2Si_2O_8 + 6H_2O \rightleftharpoons 2H^+2[Al_2(AlSi_3)_{10}O(OH)_2]^- + 3Ca(OH)_2$$ provided

calcium is removed continually from the system, the reaction proceeds to completion and H^+-montmorillonite is left as the residue. Since H^+-montmorillonite behaves as an acid, the environment assumes a low pH. Should kaolinite be the residual product, the somewhat lower cation exchange capacity of this mineral compared with montmorillonite means a lower sorption of hydrogen ions and a higher pH value for the environment.

It would appear on the basis of the above discussion that the pH of the environment is not an independent variable in chemical weathering but rather is a function of several interrelated factors: (a) the composition and structure of the parent minerals, (b) the rate of leaching of the bases, and (c) the nature and, in particular, the cation exchange capacity of the residual mineral products.

The influence of the parent mineral on the environmental pH has been discussed earlier (p. 29) in connection with abrasion pH values and does not warrant further elaboration here.

Concerning the other factors, Jenny and Leonard (1934) have demonstrated the relationship between the environmental pH and the degree of leaching in a series of soils developed on loess. In Fig. 19 the cation exchange capacity,

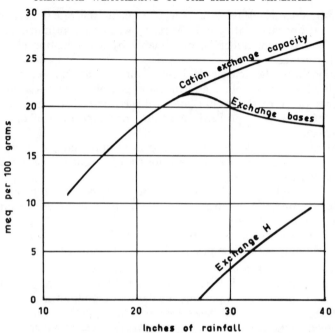

Figure 19.   Relationship between annual rainfall in inches and cation exchange capacity and base saturation in zonal soils from the United States (after Jenny and Leonard, 1934; used by permission of The Williams and Wilkins Co.).

exchangeable hydrogen ions, and pH of the surface layers of the soils are plotted against rainfall. It will be observed that whereas the cation exchange capacity increases with rainfall (apparently owing to increasing breakdown of the parent material to form clay minerals), the pH decreases, and where the rainfall is in excess of 25 inches per year, the soils become distinctly acid. This is attributed to the leaching of the exchangeable bases and their replacement at the exchange sites by hydrogen ions.

However, Jenny and Leonard's diagram shows only part of the general trend. Consider, for example, a weathering sequence developed on basalt as occurs in parts of the New England area of New South Wales. Montmorillonite is the initial weathered product but, as leaching increases, this mineral is rendered unstable through the loss of soluble constituents and is replaced by kaolinite. Ultimately destruction of the kaolinite through desilicification results in a residue of bauxite minerals. As shown diagrammatically in Fig. 20, the cation exchange capacity of the soil rises to a maximum where montmorillonite is the dominant mineral but, with the increasing development of kaolinite and bauxite, it decreases gradually toward zero. The pH curve, on the other hand, shows a reverse trend. Commencing on the alkaline side, it becomes progressively more acid through leaching of the bases and reaches a minimum at the point of conversion of montmorillonite to kaolinite. From

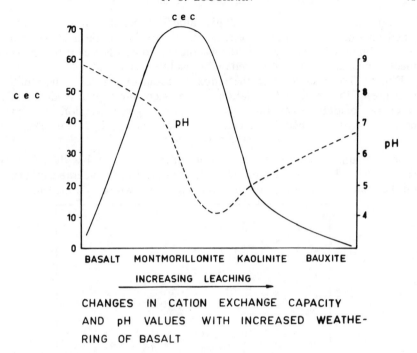

Figure 20.　Changes in cation exchange capacity (cec) and pH with increased weathering of basalt.

there, the reduced sorption by kaolinite and bauxite minerals causes the return of the curve toward neutrality.

Discussion up to the present has been based on the assumption that the entire sorption of bases and hydrogen ions is due to the inorganic constituents, that is, the clay minerals. Soil organic matter, however, is also capable of adsorbing hydrogen ions and imparting a low pH to the environment. Consequently, where organic matter is present, the acidity of the weathering environment may be independent of the residual mineralogy. Thus, organic-rich horizons in podsolic soils frequently have pH values below 4 (see Fig. 61) despite the fact that kaolinite, and not montmorillonite, is the dominant inorganic consitutent present.

Influence of Eh on the Mobilities of the Common Cations

Certain elements, in particular iron and titanium, are capable of existing in more than one valence state in combination with anions. If a difference in solubility exists between these valence states, the prevailing redox potential or Eh may seriously affect the mobility of the element. As shown in Fig. 15, precipitation of ferric hydroxide occurs at pH 3 and, under normal weathering conditions, is retained within the environment. Ferrous hydroxide, on the other hand, requires a much higher pH value for precipitation. Consequently, where conditions are suitably reducing and the pH lies on the acid side, much of the iron may be leached from the weathering zone, whereas if the environment is oxidizing, iron is stabilized in the insoluble ferric state.

Degens (1965) has defined the redox potential or Eh "as a quantitative measure of the energy of oxidation or the electron-escaping tendency of a reversible oxidation-reduction system." For a given reaction the redox potential is determined by reference to the standard hydrogen electrode

$$H_2 = 2H^+ + 2e^-$$

which, for unit activity of the hydrogen ion (i.e., pH=0), has the arbitrary value of 0.00. The hydrogen ion concentration exerts a considerable influence on the redox potential for any reaction. This is shown in Fig. 21, where the

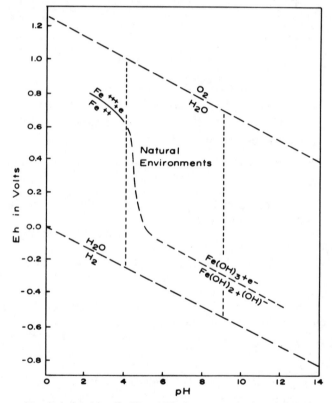

Figure 21.   Relationship of pH and Eh for some reactions involving iron.

relations between redox potential and pH are given for a number of reactions involving iron. It will be observed that the redox potential decreases with increasing pH and hence oxidation of iron proceeds more readily in alkaline environments.

Chemical reactions taking place in aqueous media are theoretically limited to those with redox potentials in the range 0.00–1.23 volt for unit activity of the hydrogen ion (pH=0).

$$H_2O = \tfrac{1}{2}O_2 + 2H^+ + 2e^- \qquad E_0 = 1.23 \text{ volt}$$
$$2H^+ + 2e^- = H_2 \qquad E_0 = 0.00 \text{ volt}$$

If the value of 1.23 volt is exceeded the oxidized form of the couple liberates oxygen from water, whereas for negative values, the reduced form decomposes water with the evolution of hydrogen. In the normal pH range of 4 to 9, the limiting values for oxidation reactions taking place in the presence of water are −0.53 and +0.99 volt (see Fig. 21).

However, little data are available on the actual redox potentials of natural weathering environments. What has been published appears confusing, for much depends on the manner in which measurements are made. Thus, Brown (1934) determined the Eh of a typical podsol and a typical gray-brown forest soil by examination of samples in the laboratory and obtained values ranging from 0.484 to 0.638 volt. Pierce (1953), on the other hand, took measurements *in situ* for a succession of soils developed on till in Wisconsin and obtained much lower values (Fig. 22).

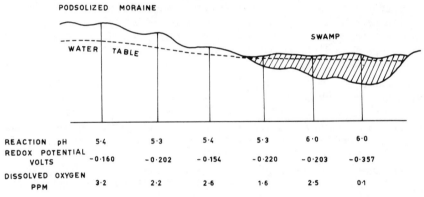

REACTION pH	5·4	5·3	5·4	5·3	6·0	6·0
REDOX POTENTIAL VOLTS	−0·160	−0·202	−0·154	−0·220	−0·203	−0·357
DISSOLVED OXYGEN PPM	3·2	2·2	2·6	1·6	2·5	0·1

Figure 22. Characteristics of ground water on a transect through a podsolized moraine and adjacent swamp region (after Pierce, 1953; reproduced from *Proc. Soil Sci. Soc. Am.* **17**, 63, by permission of the publisher).

The principal factors controlling the redox potential in weathering environments are (a) the accessibility of atmospheric oxygen and (b) the presence or absence of organic matter. Oxidation is an exothermic reaction which tends to proceed spontaneously above the zone of permanent water saturation. Below this level reducing conditions generally prevail. However, organic matter, by virtue of its ready oxidation to carbon dioxide, is a powerful reducing agent and, where present in abundance, may create a reducing environment even above the water table. It would appear, therefore, that the Eh is dependent on the climate and topography. A hot, well-drained environment favors oxidation through the rapid destruction of organic matter and the considerable lowering of the water table, whereas cool, poorly-drained environments promote accumulations of organic matter and reducing conditions.

As shown in Fig. 21, the oxidation potential for the transition of ferrous to ferric iron falls within the range anticipated for natural environments and consequently both states are common. Since a considerable difference in solubility exists between the two, the mobility of iron is greatly influenced by the prevailing Eh of the weathering environment. This was demonstrated in a simple but effective experiment carried out by Beadle and Burgess (1953). They allowed a test tube containing a soil saturated with water and to which 2% glucose solution was added to stand in the sun for a period of 3 months. Water was added periodically to offset evaporation. A scum of ferric oxide developed at the surface after a few days and at the end of the 3-month period most of the iron in the original soil was concentrated at the surface, presumably in the ferric state. The remainder of the soil developed a "bleached" appearance characteristic of the mottled zone of laterites (p. 129). This experiment not only demonstrates the ease with which iron can be oxidized but also bears witness to its mobility in the reduced ferrous state.

Titanium is another of the common rock-forming metallic elements which is capable of existing in two or more valency states. As the dioxide it is precipitated at pH values above 2.5 (Fig. 15) and hence is an immobile constituent in virtually all weathering environments. However, Sherman (1952) and Craig and Loughnan (1964) have shown that under certain conditions titanium may exhibit definite mobility.

In a study of the weathering of basalts in Hawaii, Sherman (1952) found marked accumulations (up to 25%) of titanium dioxide in the iron-enriched surface horizon of many of the soils (Table 23). The greatest accumulations occur in the humic ferruginous latosols which form under alternating wet and dry seasons adjacent to tropical rain forests. Sherman considered mobility of titanium to be possible if preceded by reduction from Ti^{4+} to a lower valence state, but according to Codell (1959), $Ti(OH)_3$ has a solubility product of 10^{-40} and hence is less soluble than alumina, while the sesquioxide, Ti_2O_3, is unstable in water, forming the dioxide, TiO_2, and evolving hydrogen.

In accounting for the movement of titanium relative to alumina in a number of weathered sequences developed on basalt in New South Wales, Craig and Loughnan (1964) drew attention to the differences in solubilities between alumina and the dioxide and hydroxide of titanium. Whereas titanium dioxide (TiO_2) is insoluble at pH values above 2.5 and alumina above 4, $Ti(OH)_4$ is not precipitated until a pH value of 5 is reached (Fig. 15). Consequently if the titanium is released as $Ti(OH)_4$ from the parent minerals of the basalt which are mainly titaniferous augite and magnetite, and the environmental pH lies between 4 and 5, movement of titanium relative to alumina within the weathered ed zone is possible. However, dehydration of $Ti(OH)_4$ to form one of the crystalline polymorphs of TiO_2 results in its immediate precipitation. On the other hand, if the titanium is present in the parent rock as one cf the crystal-

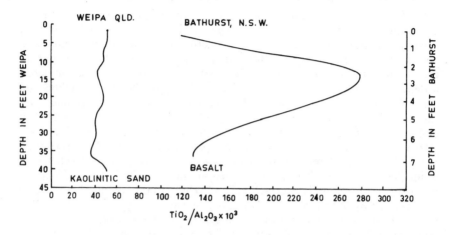

Figure 23. TiO$_2$-to-Al$_2$O$_3$ ratios for a lateritic ba.ixite developed on a kaolinitic sand at Weipa, Queensland and a red podsolic soil developed on basalt at Bathurst, New South Wales.

line forms of the dioxide (anatase or rutile), it is unaffected by the weathering processes and remains immobile. In Fig. 23 the TiO$_2$-to-Al$_2$O$_3$ ratio is plotted against depth for two weathered sequences. The one from Weipa, Queensland is developed on a kaolinitic sand in which the titanium is present as anatase, while the other, from Bathurst, New South Wales, has been derived from basalt in which the titanium is present in the pyroxene and magnetite. Differential movement of titanium relative to alumina is not apparent in the Weipa sequence whereas the Bathurst sequence shows a marked concentration of the oxide at a depth of 2–3 feet from the surface. Apparently, the differential movement of titanium at Bathurst has resulted from the release of the element from the parent rock as Ti(OH)$_4$.

The Influence of Fixation on the Mobilities of the Common Cations

It has been known since the latter part of the past century that potassium ions in solution react with many soils in an unusual manner. In contrast to most other cations, potassium tends to be retained, that is, become "fixed," by the soil and cannot be recovered with boiling in hydrochloric acid (van der Marel, 1954). The phenomenon of fixation is also apparent in many weathering reactions. Thus, despite the fact that the primary rocks contain potassium and sodium in approximately equal amounts (according to Clarke, 1924, the average igneous rock contains 3.13%K$_2$O and 3.89%Na$_2$O), the concentration of potassium in sea water is only about one-tenth that of sodium. The reason for the discrepancy is that, unlike sodium, potassium released from the primary rocks during weathering often becomes entrapped in the secondary clay products. The widespread distribution of illites in sedimentary rocks bears testimony to the prevalence of these reactions.

Over the past two decades much literature on potassium fixation has appeared, particularly in respect to soils, and it is now well established that expandable lattice clay minerals, such as vermiculites, montmorillonites, and degraded chlorites, micas, and illites, are primarily responsible for the pheno- menon. According to Barshad (1954), fixation of potassium by clay minerals is dependent on the layer charge or charge density of the mineral. Illites and chlorites which have been stripped or partially stripped of their interlayer cations, that is, are degraded, possess high charge densities and have a marked ability to fix potassium ions. On the other hand, montmorillonites have relatively low charge densities and are not particularly effective in this respect. Barshad also considered that the actual seat of the charge, that is, whether it is located in the tetrahedral or octahedral sheet of the layered silicate, is relatively unimportant. However, Wear and White (1951) compared the fixation of potassium by two montmorillonites, one in which the charge was located in the octahedral sheet and the other having a predominance of the charge in the tetrahedral sheet, and found fixation to be considerably greater in the latter.

Wear and White (1951) also investigated the reason for the preferential fixation of potassium relative to other alkali ions. Assuming each of the alkali ions in turn occupies the interlayer position in montmorillonite and that it is in 14-fold coordination with oxygen, they calculated the maximum value for the oxygen radius necessary to yield a stable configuration. They recognized that if the calculated radius for oxygen departs appreciably from the observed value of 1.40Å for oxygen in layered silicates, a stable structure would not be possible. The degree of fixation of the various alkali ions was then determined. As shown in Fig. 24, where potassium occupies the interlayer position the calculated and observed values for the oxygen radius are almost identical; significantly, maximum fixation was obtained with this ion. In brief, the unique size of the potassium ion is responsible for its preferential fixation by layered-lattice silicates.

Bassett (1960) approached the problem from a different aspect. He pointed out that the susceptibility of the micas to alteration is a function of potassium fixation, that is, it depends on how tightly the potassium ions are bound in the structure. Using infrared techniques, he determined the orientation of the O–H bond of the hydroxyl groups of both the dioctahedral and trioctahedral micas. In the trioctahedral group (biotite and phlogopite) the bond direction appears normal to the cleavage plane. This means that the hydrogen of the hydroxyl group is in juxtaposition with the potassium and consequently strain is set up through the mutual repulsion of the two positively charged ions. In the dioctahedral muscovite structure, the O–H bond is inclined to the cleavage plane and the potassium bonds to the oxygen of the hydroxyl group. In the latter case potassium assumes its preferred 14-fold coordination, made up of six oxygens forming the hexagonal cavity and one hydroxyl

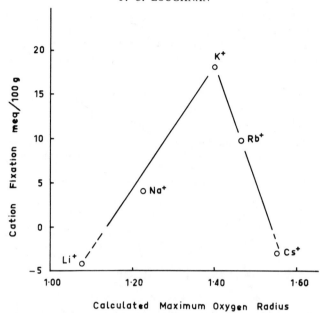

Figure 24.   Relationship between the calculated maximum value for the oxygen radius in 14-fold coordination with the alkali ions and the degree of fixation of the alkali ions (after Wear and White, 1951; used by permission of The Williams and Wilkins Co.).

group from each of the adjacent layers. However, the refinement of the muscovite structure by Radoslovich (1960) indicates that the hydroxyl-potassium bond distance is too great for the hydroxyl group to exert influence on the potassium. Consequently, doubt is cast on the validity of Bassett's explanation of the fixation mechanism.

McKenzie (1963) investigated the fixation by montmorillonites of a wide array of monovalent and divalent cations. He found that certain cations, K^+, Rb^+, and Pb^{++} for example, may be rendered nonexchangeable if the clay containing them is dried even without the application of heat. Moreover, potassium, rubidium, and ions with a radius less than 0.85 Å are significantly fixed in the montmorillonite lattice if subjected to alternate wetting and heating. In the latter case, fixation occurred irrespective of the charge on the ion.

From the above discussion, it is apparent that although the mechanism is incompletely understood, fixation of cations, particularly potassium, by clay mineral structures is prevalent and the loss of these ions from the weathering environment is retarded accordingly. Closely related to the phenomenon of fixation is the remarkable resistance shown by many potassium-bearing silicates, such as muscovite and potash feldspar, to chemical breakdown.

The Influence of Chelation on the Mobilities of the Common Cations

Chelation has been defined by Lehman (1963) "as the equilibrium reaction between a metal ion and a complexing agent, characterized by the formation of more than one bond between the metal and the molecule of the complexing agent and resulting in the formation of a ring structure incorporating the metal ion." Probably the best known chelating agent in analytical chemistry is ethylenediaminetetraacetic acid, or EDTA, which has the following structural formula:

$$HOOCCH_2\diagdown \qquad \diagup CH_2COOH$$
$$NCH_2 \cdot CH_2N$$
$$HOOCCH_2\diagup \qquad \diagdown CH_2COOH$$

With the exception of the alkalies, most metallic ions (M^{++}) react with EDTA through replacement of H^+ and the formation of complex stable ring structures.

The solubility of EDTA in water is limited but increases appreciably on complexing with metallic ions.

The effectiveness of chelating agents in accelerating weathering reactions was demonstrated by Wright and Schnitzer (1963). They leached a calcareous parent material that was essentially devoid of organic matter with EDTA and obtained a "profile" which resembled that of a podsolic soil with a bleached A_2 horizon and a sesquioxide-enriched B horizon (see Chapter VI for a description of podsolic soils). They explained the mechanism as follows: "At first decomposition of carbonates occurred. As a result, calcium and magnesium were mobilized and moved downward probably both in chelate and ionic forms. As the pH and concentration of alkaline earth metals in solution decreased at the top of the column, increasing amounts of iron and aluminum were chelated and carried downward. With depth, increasing competition from hydroxyl ions with rising pH and greater concentration of the alkaline earths caused displacement of some iron and aluminum and their precipitation as hydrated oxides."

Natural weathering environments generally contain some organic matter, representing both root secretion and the accumulation of decaying plant debris, and it is probable that components of this organic matter are capable of complexing metallic ions in much the same manner as EDTA. If this is the case, organic matter could play an important role in the weathering of the silicates through the direct removal of metallic ions from the mineral, thus initiating breakdown, or through the effective mobilization of otherwise immobile metallic ions.

Investigations carried out over the past decade or so (Wallace *et al.*, 1955;

Himes and Barber, 1952; Schatz *et al.*, 1957; DeKock, 1960; and many others) have established that at least some soil organic matter possesses chelating properties. but the evidence is still little better than circumstantial and a great deal more work will be necessary before the extent of this type of activity and its full influence on chemical weathering are known.

The Role of Leaching in Chemical Weathering

Undoubtedly the most important single factor controlling the rate of breakdown of parent minerals and the genesis of specific secondary products is the quantity of water leaching through the weathering environment. Repeated flushings by rainwater tend to remove the soluble constituents released by the hydrolyzing processes at the mineral surfaces and permit the weathering reactions to proceed toward completion. The soluble constituents move downward through the weathering zone and ultimately, by way of subsurface drainage, reach rivers, lakes, or the ocean. The role of leaching in weathering and soil formation was fully appreciated by Jenny (1941) when he wrote, "The most active agency in soil profile development is percolating water. As long as water passes through the solum, substances are dissolved, translocated, precipitated and flocculated and the soil is not in a state of rest."

Given sufficient rainfall, permeability, and time, without erosion or geological disturbance, even the most stable parent minerals, such as quartz, can be destroyed because all materials have some degree of solubility. The lateritic bauxites near Weipa in North Queensland (Loughnan and Bayliss, 1961) present an excellent example of this. Here, a kaolinitic sandstone, consisting of approximately 90% quartz and 10% kaolinite, has been intensively leached under a tropical monsoonal climate and over a considerable period of time to yield a residual assemblage of bauxite and laterite minerals with a small amount of kaolinite and only about 5% quartz (Fig. 55). The quartz persisting in the highly leached zone shows evidence of chemical attack and is generally coated by bauxite and laterite minerals. Complete destruction of the quartz apparently has been prevented by these protective coatings.

Pedro (1961) has demonstrated the effectiveness of leaching by distilled water in a series of remarkable laboratory experiments designed to simulate weathering under humid, tropical conditions. Broken fragments of igneous rocks, 10 to 20 grams in weight, were placed in the tube of a Soxhlet extractor, modified for the purpose as shown in Fig. 25, and distilled water was continuously recycled at the rate of 3 liters per day (calculated as the equivalent to a daily rainfall of 39.4 inches) by maintaining the distilling flask at boiling point. The actual temperature of the water percolating through the rock fragments was in the vicinity of 65°C. By means of a siphon on the return line, two different environments were created within the tube, (a) an upper, non-inundated "atmospheric zone" and (b) a periodically saturated zone of "fluctuating water table." The leaching extended over a period of 2 years and several rock types were examined.

The volcanic rocks, basalt and trachyandesite, yielded the most spectacular results. In the "atmospheric zone" these rocks developed an ocher crust

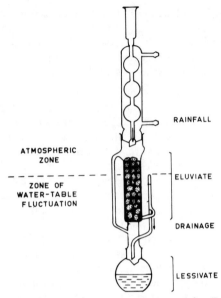

Figure 25.   Experimental weathering apparatus (after Pedro, 1961).

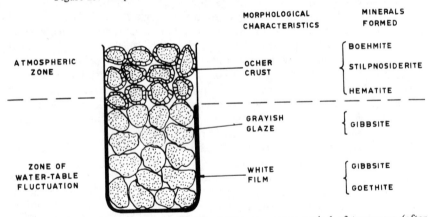

Figure 26.   Results of leaching of basaltic fragments over a period of two years (after Pedro, 1961).

composed of boehmite, stilpnosiderite (limonite), and hematite whereas in the zone of "fluctuating water table," a gibbsite glaze appeared on the rock fragments and, in association with goethite, formed a film on the inside of the tube (Fig. 26). Material leached from the rock fragments was collected in the flask and subsequent examination revealed a predominance of silica with alkalies, alkaline earths, and a little alumina. The final pH of the solution lay between 9 and 10. Apparently, the solutions returning to the flask were not filtered and probably some colloidal material was carried over to the flask.

Leaching of granite did not produce the same marked visual effects although analyses of the leachate revealed the removal of appreciable quantities of silica, alkalies, and alkaline earths from the rock.

Goldschmidt (1937) suggested that the behavior of the various ions in weathering reactions can be related directly to their ionic potential, a concept originally introduced by Cartledge (1928). Ionic potential is a fundamental property of an element, which is closely related to its electronegativity (Pauling, 1960), and is expressed numerically as the ratio of the charge in valency units (Z) to the ionic radius in Ångström units (r). As shown in Fig. 27, ions can be arranged into three groups: (a) those with a low ionic potential ($Z/r < 3.0$) such as Na^+, K^+, Ca^{++}, and Mg^{++}, which tend to pass into true

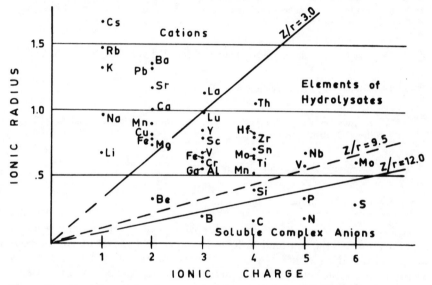

Figure 27.   Grouping of the elements according to their ionic potentials (after Gordon and Tracey, 1952).

ionic solution during the weathering processes, (b) those with intermediate ionic potentials (Z/r lies between 3.0 and 9.5), such as Al^{3+}, Fe^{3+}, and Ti^{4+}, which are precipitated and become concentrated in the residue, and (c) those with high ionic potentials ($Z/r > 9.5$) including silicon, nitrogen, and phosphorus, which form soluble anionic radicals on weathering. In general, ionic potential is a useful guide to the behavior of the various ions. Nevertheless, it should be borne in mind that other factors such as the parent mineral structure, the environmental pH and Eh, and so on may exert an equally important influence in controlling the loss or retention of a specific ion.

Conclusions

From this brief consideration of the chemical factors involved, it is apparent that, although fixation, chelation, and variations in pH and Eh may play an important role, the extent of chemical weathering and the nature of the residual products are primarily functions of the degree of leaching to which the parent rock has been subjected. Provided the pH of the environment falls within the normal range of 4 to 9, it is possible to draw conclusions concerning the loss or retention of the common metallic ions in relation to variations in these factors (Table 6).

TABLE 6
Mobilities of the Common Cations

1.	Ca^{++}, Mg^{++}, Na^+—readily lost under leaching conditions
2.	K^+—readily lost under leaching conditions but rate may be retarded through fixation in the illite structure
3.	Fe^{++}—rate of loss dependent on the redox potential and degree of leaching
4.	Si^{4+}—slowly lost under leaching conditions
5.	Ti^{4+}—may show limited mobility if released from the parent mineral as $Ti(OH)_4$; if in the TiO_2 form, immobile
6.	Fe^{3+}—immobile under oxidizing conditions
7.	Al^{3+}—immobile in the pH range of 4.5–9.5

Left margin: Increasing rate of loss from the environment (arrow pointing upward)

(1) The behavior of aluminum is independent of all factors and residual concentrations of the ion tend to form through the removal of mobile constituents from the parent material. However, in exceptional environments where the pH values lie outside the range of 4 to 9, solution and loss of aluminum are possible.

(2) Titanium, if present in the parent material as one of the crystalline oxides, rutile or anatase, behaves in a similar manner to aluminum but, if released from the parent material as $Ti(OH)_4$ through the breakdown of titanium-bearing silicates, such as titanaugite, or oxides, such as ilmenite or titaniferous magnetite, it may exhibit limited mobility provided the environmental pH is below 5. Dehydroxylation of $Ti(OH)_4$ to one of the crystalline oxides renders titanium immobile.

(3) The behavior of iron is complicated by the interrelationship of pH and Eh. Oxidation stabilizes iron in the insoluble ferric state in which case it does not tend to form silicates but rather persists as either the oxide or the hydroxide. Reduction, on the other hand, renders it potentially mobile in the ferrous form probably as the bicarbonate. The sulfides, pyrite and marcasite, are rare as products of weathering.

(4) The loss or retention of the alkalies, magnesium, calcium, and ferrous

iron is essentially a function of the degree of leaching, although fixation of potassium and possibly magnesium by certain silicate structures, may retard solution of these constituents.

(5) Silicon, either present in the parent material in the amorphous state or released from silicate minerals of the parent rock, has a low but constant solubility. Quartz, on the other hand, has a solubility of only about one-tenth that of amorphous silica (Krauskopf, 1959) and, hence, is considerably more stable. Alumina in the amorphous state or as gibbsite appears to have a marked affinity for silica and presumably reaction between these constituents results in the crystallization of kaolinite or halloysite. Because of this affinity, desilicification of kaolins to yield bauxites occurs only under intense leaching conditions.

The Influence of Crystal Structure

It has been noted previously that all minerals do not weather with the same ease. Some are rapidly destroyed in the weathering environment whereas others appear little affected and ultimately are transported by running water, wind, and so on to be deposited as detrital grains. That the crystal structure must influence the rate of decay of minerals is apparent from the fact that some minerals containing highly mobile alkali ions as essential constituents (e.g., muscovite, potash, and soda feldspars) can exhibit remarkable stability to the weathering solutions. There have been many attempts to determine the precise role played by crystal structures in the breakdown processes and to establish a stability sequence for the various minerals.

Goldich (1938) was one of the earliest of the modern workers in the field. From a detailed study of the weathering of granite gneiss, diabase, and amphibolite he concluded that the weathering sequence for the common rock-forming minerals is in the following order.

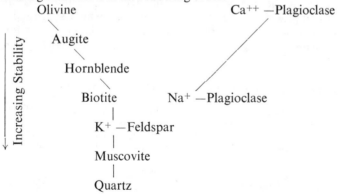

The order coincides exactly with that of Bowen (1928) for the crystallization of minerals from a silicate melt and, if muscovite and the plagioclases are disregarded, with the classification of silicate structures based on the increasing number of Si–O–Si bonds from zero in the case of olivine (nesosilicates) to 4 in the case of quartz (tectosilicate).

The significance of Goldich's conclusions was appreciated by Keller (1954). Using the approximate values for the "energies of formation of cation-oxygen bonds in silicate minerals and glasses from a reference state of gaseous ions" given by Huggins and Sun (1946), he calculated the bond energies for the common oxides (Table 7). Then, assuming a constant number of oxygens

TABLE 7

Molal Energies of Formation of Corresponding Oxides in Silicate Glasses and Minerals[a]

Ion, M	em(kg cal)
Ca^{++}	839
Mg^{++}	912
Fe^{++}	919
Na^+	322
K^+	299
H^+ (in OH)	515
Ti^{4+}	2882
Al^{3+} (in aluminates)	1878
(in Al silicates)	1793
Si^{4+}	
in $MSiO_4$	3142
in MSi_2O_7	3137
in $MSiO_3$	3131
in MSi_4O_{11}	3127
in MSi_2O_5	3123
in $MSiO_2$	3110

[a] After Keller (1954).

in each structure (24 was found convenient), he computed the "energies of formation" for each of the common silicate linkages. He found these increased with the complexity of the structure (Table 8), which suggested a possible correlation between the weathering stability and the energy of formation of the silicate minerals. However, when the cation linkages were

TABLE 8

Energy Sequence from Nesosilicate to Tectosilicate, 24 Oxygens[a]

Type	Factor	Adjusted bonding energy = 24 O's
Nesosilicate	$6 \cdot SiO_4$	18,852 kg cal + 12 M^{++} bridges
Sorosilicate	$\frac{24}{7} 2 \cdot Si_2O_7$	21,511 kg cal + 10$\frac{2}{7}$ M^{++} bridges
Inosilicate	$8 \cdot SiO_3$	25,048 kg cal + 8 M^{++} bridges
Inosilicate	$\frac{24}{11} 4 \cdot Si_4O_{11}$	27,290 kg cal + 7 M^{++} bridges and 2H (commonly)
Phyllosilicate	$\frac{24}{5} 2 \cdot Si_2O_5$	29,981 kg cal + 6 M^{++} bridges and 4H (commonly)
Tectosilicate	$12 \cdot SiO_2$	37,320 kg cal + no M^{++} bridges

[a] After Keller (1954).

TABLE 9

Common Minerals and Their Bonding Energies[a]

Mineral	Formula	Bonding energy, 24 O's (kg cal)
Gehlenite	$Ca_2Al(AlSi)O_7$	26,890
Spinel	$MgAl_2O_4$	28,008
Forsterite	Mg_2SiO_4	29,796
Fayalite	Fe_2SiO_4	29,800
Epidote	$Ca_2(Al_2Fe)Si_3O_{12}(OH)$	30,020
Idocrase	$Ca_{10}(MgFe)_2Al_4Si_9O_{34}(OH)_4$	30,360
Äkermanite	$Ca_2MgSi_2O_7$	30,391
Biotite	$K(MgFe)_3(AlSi_3)O_{10}(OH)_2$	30,475
Augite	$Ca(MgFeAl)(AlSi)_2O_6$	30,728
Topaz	$Al_2SiO_4(FOH)_2$	31,712
Staurolite	$Fe_2Al_9O_7Si_4O_{16}(OH)$	31,823
Nepheline	$NaAlSiO_4$	31,860
Almandine	$Fe_3Al_2Si_3O_{12}$	31,878
Hornblende	$(CaNa)_2(MgFeAl)_5(AlSi)_8O_{22}(OH)_2$	31,883
Anorthite	$CaAl_2Si_2O_8$	31,935
Diopside	$CaMgSi_2O_6$	32,052
Dickite	$Al_2Si_2O_5(OH)_4$	32,165
Tremolite	$Ca_2Mg_5Si_8O_{22}(OH)_2$	32,284
Enstatite	$MgSiO_3$	32,344
Analcite	$NaAlSi_2O_5(OH)_2$	32,410
Muscovite	$KAl_2(AlSi_3)O_{10}(OH)_2$	32,494
Talc	$Mg_3Si_4O_{10}(OH)_2$	32,516
Pyrophyllite	$Al_2Si_4O_{10}(OH)_2$	32,558
Sillimanite	Al_2SiO_5	32,957
Orthoclase	$KAlSi_3O_8$	34,266
Albite	$NaAlSi_3O_8$	34,335
Quartz	SiO_2	37,320

[a] After Keller (1954).

taken into account, the order changed and correlation with the Goldich sequence broke down (Table 9).

Gruner (1950) attempted to establish a stability sequence for the silicate minerals based on both structure and composition. Using Pauling's data for the electronegativities of the elements, he first calculated the average negativity of the cations present in each mineral and then multiplied this value by an empirically determined factor, the "bridging factor," which he believed accounted for variations in the mineral structures. The product was termed the "energy index of the mineral" and the values for some of the silicates are given in Table 10. It will be observed that the "energy indices" are in the order of stability observed by Goldich, quartz having the maximum and olivine the minimum value. However, few students of weathering would agree with the positions of some of the less common silicate minerals. For example, chlorites rank among the more unstable of the phyllosilicates, yet they have an "energy index" greater than muscovite. Again, on the values obtained, montmorillonite, hydromicas, and vermiculite appear more stable than zircon.

TABLE 10

Energy Indices of Minerals[a]

Mineral	Energy index	Mineral	Energy index	Mineral	Energy index
Quartz	1.80	Tourmaline	1.55	Enstatite	1.40
Montmorillonite	1.77	Sillimanite	1.55	Almandite	1.40
Kaolinite	1.75	Kyanite	1.55	Aegerine	1.40
Pyrophyllite	1.73	Andalusite	1.55	Leucite	1.38
Stilbite	1.65	Analcite	1.53	Vesuvianite	1.36
Lawsonite	1.65	Cordierite	1.51	Sphene	1.36
Hydromicas	1.64	Albite	1.49	Diopside	1.35
Vermiculite	1.63	Glaucophane	1.49	Augite	1.35
Zircon	1.63	Orthoclase	1.48	Grossularite	1.34
Beryl	1.60	Zoisite	1.46	Nepheline	1.30
Staurolite	1.59	Epidote	1.46	Wollastonite	1.30
Serpentine	1.58	Tremolite	1.45	Andradite	1.30
Chlorites	1.58	Hornblende	1.45	Forsterite	1.28
Chloritoid	1.58	Anorthite	1.44	Fayalite	1.28
Talc	1.56	Phlogopite-		Äkermanite	1.24
Muscovite	1.55	biotite	1.42	Larnite	1.14
Topaz	1.55	Pyrope	1.42		

[a] After Gruner (1950).

Fairbairn (1943) approached the problem from a different aspect. He considered the degree of packing in minerals should have a bearing on their relative stabilities and proposed an index, defined as the ratio of the volume of ions in the unit cell to the total volume of the unit cell. The computed values for a wide range of minerals are given in Table 11. It will be noted that the values for some of the more common minerals are at variance with the Goldich sequence.

TABLE 11

Packing Indices of Minerals[a]

Mineral	Packing index	Mineral	Packing index	Mineral	Packing index
Montmorillonite	3.6	Muscovite	5.5	Forsterite	5.9
Analcite	3.7	Talc	5.6	Andalusite	6.0
Vermiculite	4.3	Mullite	5.6	Schorlite	6.0
Cordierite	4.7	Fayalite	5.7	Andradite	6.1
Microcline	5.0	Hornblende	5.7	Sillimanite	6.2
Orthoclase	5.0	Pyrophyllite	5.7	Anatase	6.3
Quartz	5.2	Kaolinite	5.8	Staurolite	6.7
Gehlenite	5.3	Dickite	5.9	Zircon	6.7
Monazite	5.4	Diopside	5.9	Kyanite	7.0
Serpentine	5.4	Augite	5.9	Diaspore	7.0
Biotite	5.5	Hypersthene	5.9		

[a] After Fairbairn (1943).

TABLE 12

Weathering Sequence of Clay-Size Minerals in Soils and Sedimentary Deposits[a]

Weathering stage and symbol	Clay-size mineral occurring at various stages of the weathering sequence
1, Gp	Gypsum (also halite, etc.)
2, Ct	Calcite (also dolomite, aragonite, etc.)
3, Hr	Olivine-hornblende (also diopside, etc.)
4, Bt	Biotite (also glauconite, chlorite, antigorite, etc.)
5, Ab	Albite (also anorthite, microcline, stilbite, etc.)
6, Qtz	Quartz (also cristobalite, etc.)
7, Il	Illite (also muscovite, sericite, etc.)
8, X	Hydrous mica—Intermediates
9, Mt	Montmorillonite (also beidellite, etc.)
10, Kl	Kaolinite (also halloysite, etc.)
11, Gb	Gibbsite (also boehmite, etc.)
12, Hm	Hematite (also goethite, limonite, etc.)
13, An	Anatase (also rutile, ilmenite, corundum, etc.)

[a] After Jackson et al. (1948); used by permission of The Williams and Wilkins Co.

Jackson and his colleagues (Jackson et al., 1948, 1952) have proposed a weathering sequence for clay-size ($<2 \mu$) minerals in soils (Table 12). They contend that the presence of certain minerals can be used as indicators of the weathering intensity to which a particular soil has been subjected and hence the concept is essentially that of a mineral facies. Thirteen stages have been recognized, and the indicative minerals in order of increasing weathering intensity are gypsum, calcite, olivine-hornblende, biotite, albite, quartz, hydrous micas, mica intermediates (degraded micas), montmorillonite, kaolinite, gibbsite, hematite, and anatase. Fieldes and Swindale (1954) believed a weakness of the scheme is that "it fails to make clear the way in which the various secondary minerals are related to the primary minerals from which they are formed." They have suggested certain modifications (Table 13).

In the investigation of the distribution of heavy minerals in sedimentary rocks, Pettijohn (1941) compared the frequency of occurrence of each species in Recent sediments with its average frequency of occurrence in non-Recent sediments and established an order of persistence (Table 14). Numerical values were assigned to the various species ranging from -3 in the case of anatase, the most stable, to 22 in the case of olivine, the least stable. Pettijohn argued that, since the persistence of a mineral is a function of its resistance to chemical attack, the order given must approximate to a weathering stability sequence. Although the physical and chemical conditions prevailing in deeply buried sediments are vastly different from those encountered in weathering (in fact, authigenesis, a process somewhat opposed to weathering, is favored by burial), nevertheless Pettijohn's order for the common rock-forming minerals is in agreement with the Goldich series.

Reiche (1943) has devised a "weathering potential index" based on the following empirical formula:

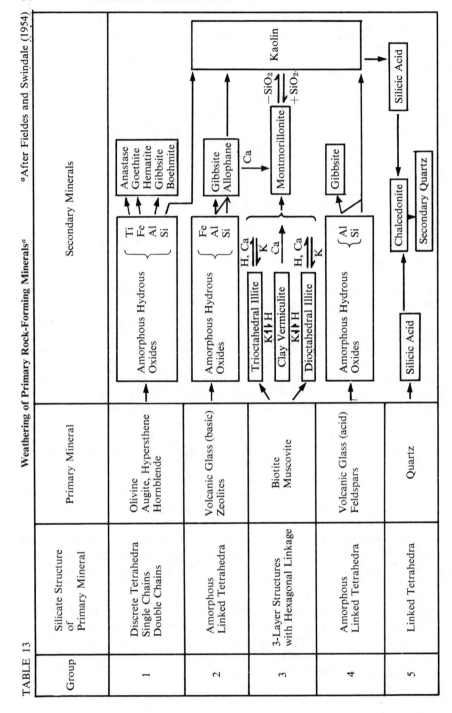

TABLE 13

Weathering of Primary Rock-Forming Minerals[a]

[a]After Fieldes and Swindale (1954)

TABLE 14
Persistence Order of Minerals[a, b]

−3.	Anatase	10.	Kyanite
−2.	*Muscovite*	11.	Epidote
−1.	Rutile	12.	*Hornblende*
1.	Zircon	13.	Andalusite
2.	Tourmaline	14.	Topaz
3.	Monazite	15.	Sphene
4.	Garnet	16.	Zoisite
5.	*Biotite*	17.	*Augite*
6.	Apatite	18.	Sillimanite
7.	Ilmenite	19.	Hypersthene
8.	Magnetite	20.	Diopside
9.	Staurolite	21.	Actinolite
		22.	*Olivine*

[a] After Pettijohn (1941); used by permission of The University of Chicago Press.
[b] Italics signify common minerals listed in the Goldich sequence.

$$\text{Weathering potential index} = \frac{100 \times \text{moles}(Na_2O + K_2O + CaO + MgO - H_2O)}{\text{moles}(Na_2O + K_2O + CaO + MgO + SiO_2 + Al_2O_3 + Fe_2O_3)}$$

Although the formula is primarily concerned with chemical composition, structure is indirectly taken into account as shown in Table 15, where the

TABLE 15
Weathering Potential Indices for a Number of Magnesium Silicates

Mg_2SiO_4	Forsterite	66
$Mg_3Si_2O_7$	Not known	60
$MgSiO_3$	Enstatite	50
$Mg_7(Si_4O_{11})_2(OH)_2$	Anthophyllite	40
$Mg_3Si_4O_{10}(OH)_2$	Talc	29

weathering potential has been calculated for a series of magnesium silicates representing different structural groups. It will be observed that there is a concomitant decrease in the indices with increasing complexity of the silicate structure. The index also reflects variations in composition within a particular structural group such as the plagioclases. However, when a wide array of minerals is examined (Table 16) certain anomalies appear. Thus analcite, one of the least stable of the silicate minerals to chemical weathering, has an index less than that of muscovite or quartz.

Marshall (1964) has offered an interesting explanation to account for the difference in weathering stability within the plagioclase series. He quoted the work of Armstrong on electrodialysis of feldspars which showed the following order of removal of constituents

$$CaO > Na_2O > K_2O > SiO_2 \gg Al_2O_3$$

from the minerals, and pointed out that where the feldspars have an Al/Si

TABLE 16

Weathering Potential Indices[a, b]

Forsterite[d]	66	Biotite[c]	22
Olivine[c]	54	Leucite[d]	17
Wollastonite[d]	50	Albite[c]	13
Enstatite[d]	50	Orthoclase[c]	12
Diopside[d]	50	Quartz[c]	0
Tremolite[d]	40	Sillimanite[d]	0
Augite[c]	39	Muscovite[c]	−10.7
Hornblende[c]	36	Analcite[d]	−17
Talc[d]	29	Pyrophyllite[d]	−20
Nepheline[d]	25	Kaolinite[d]	−67
Anorthite[d]	25	Boehmite[d]	−100
Epidote[d]	23	Gibbsite[d]	−300

[a] After Reiche (1943).
[b] Italics signify common minerals listed in the Goldich sequence.
[c] Average value obtained by Reiche.
[d] Calculated from theoretical formula.

ratio of 1 : 3 as in albite, "weakened SiO fragments which had lost Al^{3+} could, to a considerable extent, remain attached to the weathered framework beneath." However, where the ratio is 1 : 1 as in anorthite, "removal of Al^{3+} would, on the average, cause breakdown to very small incoherent fragments."

Loughnan (1962) has emphasized the role played by mineral structures in controlling access of the percolating waters to the soluble cations. Olivine, for example, consists of isolated silica tetrahedra bonded together by magnesium and ferrous ions. Both these cations are readily leached from the margins and fractured surfaces of the crystals and the loss of the cationic bridges releases the tetrahedra. Fresh surfaces are thereby exposed for further attack by the percolating waters. Consequently, the chemical weathering of olivine proceeds rapidly. The alkali feldspars, on the other hand, are relatively resistant to breakdown despite the fact that highly mobile potassium and sodium ions form essential constituents, because the framework structure of the tetrahedra hinders escape of the alkali ions. To free these mobile cations from the feldspars, the tetrahedral chains must be ruptured. Apparently the Al–O–Si bonds form the weakest links in the chains and the greater the substitution of aluminum for silicon in the tetrahedra the greater is the number of these weak links. This probably accounts for the considerable range in weathering stability within the feldspar group.

From the above discussion it is apparent that the precise role played by the crystal structure in determining the rate of decay of a specific mineral is still far from understood. Whereas the Goldich stability sequence for the common rock-forming minerals seems well established, an entirely satisfactory explanation has not yet been advanced to account for the similarities between this sequence and Bowen's reaction series. Moreover, the apparent correlation between the weathering stability of a silicate mineral and the degree of polymerization of the silica tetrahedra within the mineral structure cannot be extended beyond the few rock-forming minerals listed by Goldich. Zircon and andalusite, for example, are resistant to chemical breakdown, yet both are members of the nesosilicates which, like olivine, do not contain polymerized tetrahedra. Again, the marked disparity in weathering stability between albite and anorthite, both of which are tectosilicates, and between the phyllosilicates, muscovite and biotite, suggests that this apparent correlation is somewhat fortuitous.

THE SOLUBLE PRODUCTS OF CHEMICAL WEATHERING

Solution is essential to chemical weathering. Minerals break down primarily because some of their constituent atoms and ions are dissolved and effectively removed from the environment. The loss of these constituents renders the existing mineral structures unstable and new crystalline phases tend to form in their stead. Under natural conditions, solution is generally implemented by downward moving meteoric waters which percolate through the soil and rock and, depending on the prevailing climate, topography, and rock type, often penetrate to considerable depth. During periods of aridity, some of this water may return to the surface and become lost through evaporation but most of it, particularly in humid areas, drains slowly under the influence of gravity through pores and fissures in the rocks and eventually emerges as seepages and springs farther down the valley floor. Here, mingling with surface drainage waters takes place in creeks, rivers, and lakes. Since the dissolved constituents of the subsurface water have been derived essentially from the rocks through which the water has traversed, the composition should, in general, reflect the nature and extent of chemical weathering to which the rocks of the particular area have been subjected. Regrettably, this aspect of chemical weathering has been neglected in the past and specific information is sparse.

White et al. (1963) have presented data on the soluble constituents in subsurface waters associated with various rock types. They selected approximately 150 analyses taken "largely from environments in which the waters were most likely to be atmospheric precipitation that was influenced primarily by reactions with the rocks in which they are found." As shown by the few representative examples in Table 17, the analyses, in general, reflect the composition of the parent rock although some anomalies are apparent. Rocks rich in the alkaline earths tend to yield waters with relatively high concentrations of these ions, and consequently high pH values, whereas the waters associated with acid igneous rocks contain a predominance of silica and have low pH values. A notable feature of the analyses is the very low concentrations of aluminum and iron in the ground waters, a trend which appears to be in agreement with the conclusions concerning the mobilities of these constituents, reached earlier in this chapter.

More direct evidence of the relationship between the secondary mineral products and the composition of the drainage waters has been furnished by Feth et al. (1964) in a study of the ground waters associated with granites in

TABLE 17

Chemical Analyses of Some Ground Waters Associated with Various Rock Types[a]

| Rock type | pH | Parts per million | | | | | | | | | |
		SiO$_2$	Al	Fe	Ca	Mg	Na	K	HCO$_3$	SO$_4$	Cl
Rhyolite	6.6	37	0.1	0.2	3.6	0.8	3.9	2.3	21	2.6	1.4
Granite	6.6	39	0.9	1.6	27	6.2	9.5	1.4	93	32	5.2
Basalt	7.7	44	0.1	0.0	54	20	51	7.2	242	61	46
Gabbro	6.7	39	0.0	5.1	5.1	2.3	6.2	3.2	37	9.2	1.0
Serpentine	8.3	31	0.2	0.1	9.5	51	4.0	2.2	276	2.6	12
Andesite	7.2	31	0.2	0.2	14	5.6	9.6	0.4	74	0.1	8.8
Syenite	7.6	19	0.0	0.1	9.5	2.3	2.8	0.6	38	2.8	2.1
Sandstone	7.4	12	0.2	0.1	50	6.0	2.4	3.0	184	2.1	1.8
Arkose	6.7	35	0.3	0.2	9.6	1.9	5.1[b]		38	7.4	1.8
Greywacke	8.2	12	0.0	0.6	74	20	34	1.2	381	26	2.7
Shale	8.1	17	0.1	8.2	48	29	447	8.4	579	1.5	536
Limestone	8.2	11	0.1	0.0	61	34	9	1.1	291	20	1.1
Dolomite	7.4	14	0.2	4.2	178	86	76	3.5	285	707	11
Chert	6.5	26	0.9	0.9	26	1.9	7.4	2.8	68	34	2.2
Schist	6.3	14	0.0	0.2	3.1	1.2	3.3	0.8	21	1.2	2.4
Glacial deposit	7.4	14	0.8	0.3	120	49	122	6.2	399	439	6.0

[a] Taken from White *et al.* (1963).

[b] Total Na+K.

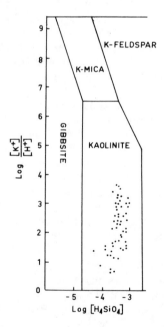

Figure 28. Stability relations of phases in the system K$_2$O–Al$_2$O$_3$–SiO$_2$–H$_2$O at 25°C and 1 atmosphere total pressure as functions of [K$^+$]/[H$^+$] and [H$_4$SiO$_4$] (after Feth *et al.* 1964).

the Sierra Nevada of western United States. Using published data for the standard free energies of formation of compounds and ions at 25°C and 1 atmosphere total pressure and by making several reasonable assumptions, they determined the stability relations of the mineral phases in the systems $Na_2O-Al_2O_3-SiO_2-H_2O$ and $K_2O-Al_2O_3-SiO_2-H_2O$. As shown in Figs. 28 and 29, the stability fields can be plotted as a function of the concentration

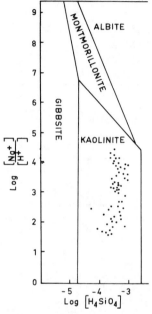

Figure 29.   Stability relations of phases in the systems $Na_2O-Al_2O_3-SiO_2-H_2O$ at 25°C and 1 atmosphere total pressure as functions of $[Na^+]/[H^+]$ and $[H_4SiO_4]$ (after Feth *et al.* 1964).

of silicic acid and of the ratio of the concentration of each of the alkali ions to that of hydrogen. The compositions of the ground waters were plotted on the same diagrams and it was found that all fell within the stability fields of kaolinite. This indicated that kaolinite is the solid phase in equilibrium with the ground waters and that the small quantities of montmorillonite and mica present in the weathered residuum are unstable and eventually will be converted to kaolinite.

Estimates of the amount of soluble material removed annually from the earth's surface by chemical weathering are lacking. However, if it be assumed that the bulk of the dissolved matter in rivers has been derived from the weathering of rocks, then the quantity is appreciable, as shown by the calculations of Livingstone (1963) which indicate a value of 3905 million metric tons per annum. Livingstone also computed the mean composition of river waters for each of the continents and the values are shown in Table 18.

TABLE 18

Mean Chemical Composition of River Waters of the World in Parts per Million[a]

Area	HCO_3	SO_4	Cl	NO_3	Ca	Mg	Na	K	Fe	SiO_2	Sum
North America	68	20	8	1	21	5	9	1.4	0.16	9	142
Europe	95	24	6.9	3.7	31.1	5.6	5.4	1.7	0.8	7.5	182
Asia	79	8.4	8.7	0.7	18.4	5.6	9.3[b]		0.01	11.7	142
Africa	43	13.5	12.1	0.8	12.5	3.8	11	—	1.3	23.2	121
South America	31	4.8	4.9	0.7	7.2	1.5	4	2	1.4	11.9	69
Australia	31.6	2.6	0	0.05	3.9	2.7	2.9	1.4	0.3	3.9	59
World	58.4	11.2	7.8	1.0	15	4.1	6.3	2.3	0.67	13.1	120
Anions[c]	0.958	0.233	0.220	0.017	—	—	—	—	—	—	1.428
Cations[c]	—	—	—	—	0.750	0.342	0.274	0.059	—	—	1.425

[a] After Livingstone (1963).

[b] Total $Na+K$.

[c] Millequivalents of strongly ionized components.

Probably the most interesting aspect of this table is the considerably greater abundance of calcium and bicarbonate ions in the river waters of the continents of the northern hemisphere compared with those of the southern hemisphere.

REFERENCES

Armstrong, L. C. (1940), Decomposition and alteration of feldspar and spodumene by water, *Am. Mineralogist* **25**, 810–820.

Bardossy, G. (1959), The geochemistry of Hungarian bauxites, *Acta Geol. Acad. Sci. Hung.* **6**, 1–47.

Barshad, I. (1954), Cation exchange in micaceous minerals. II. Replaceability of ammonium and potassium from vermiculite, biotite and montmorillonite, *Soil Sci.* **78**, 57–76.

Bassett, W. A. (1960), Role of hydroxyl orientation in mica alteration, *Bull. Geol. Soc. Am.* **71**, 449–456.

Beadle, N. C. W. and A. Burgess (1953), A further note on laterites, *Australian J. Sci.* **15**, 170–171.

Brown, L. A. (1934), Oxidation-reduction potentials in soils, *Soil Sci.* **37**, 65–76.

Bowen, N. L. (1928), *The Evolution of the Igneous Rocks*. Princeton Univ. Press, Princeton, New Jersey.

Cartledge, G. H. (1928), Studies in the periodic system, *J. Chem. Soc. Am.* **50**, 2855–2872.

Clarke, F. W. (1924), Data of geochemistry, *U.S. Geol. Surv. Bull.* **770**.

Codell, M. (1959), *Analytical Chemistry of Titanium Metals and Compounds*. Interscience, New York.

Correns, C. W. (1949), *Einfuhrung in die Mineralogie*. Springer-Verlag, Berlin.

Craig, D. C., and F. C. Loughnan (1964), Chemical and mineralogical transformations accompanying the weathering of basic volcanic rocks from New South Wales, *Australian J. Soil Res.* **2**, 218–234.

Degens, E. T. (1965), *Geochemistry of Sediments*. Prentice-Hall, Englewood Cliffs, New Jersey.

DeKock, P. C. (1960), The effects of natural and synthetic chelating substances on the mineral status of plants, *Trans. 7th Intern. Congr. Soil Sci., Wisconsin*, pp. 574–579.

DeVore, G. W. (1958). The surface chemistry of feldspars as an influence on their decomposition products, *Proc. Nat. Conf. Clays and Clay Minerals* **6**, 26–41.

Fairbairn, H. W. (1943), Packing in ionic minerals, *Bull. Geol. Soc. Am.* **54**, 1305–1374.

Feth, J. H., C. E. Robertson, and W. L. Polzer (1964), Sources of mineral constituents in water from granitic rocks, Sierra Nevada, California and Nevada, *U.S. Geol. Surv. Water Supply Paper* **1535-I**.

Fieldes, M., and L. D. Swindale (1954), Chemical weathering of silicates in soil formation, *J. Sci. Tech. New Zealand* **56**, 140–154.

Frederickson, A. F. (1951), Mechanism of weathering, *Bull. Geol. Soc. Am.* **62**, 221–232.

Garrels, R. M., and P. Howard (1951), Reactions of feldspar and mica with water at low temperatures and pressures, *Proc. Nat. Conf. Clays and Clay Minerals* **6**, 68–88.

Goldich, S. S. (1938), A study of rock weathering, *J. Geol.* **46**, 17–58.

Goldschmidt, V. M. (1937), The principles of the distribution of chemical elements in minerals and rocks, *J. Chem. Soc. London*, pp. 655–672.

Gordon, M., and J. I. Tracey (1952), Origin of the Arkansas bauxite deposits. Problems of clay and laterite genesis, *Am. Inst. Min. Met.*, pp. 12–34.

Graham, E. R. (1941a), Acid clay, an agent in chemical weathering, *J. Geol.* **49**, 392–401.

Graham, E. R. (1941b), Colloidal acids as factors in the weathering of anorthite, *Soil Sci.* **52**, 291–295.

Gruner, J. W. (1950), An attempt to arrange silicates in the order of reaction energies at relatively low temperatures. *Am. Mineralogist* **35**, 137–148.

Hay, R. L., and R. J. Moiola (1963), Authigenic silicate minerals in Searles Lake, California, *Sedimentology* **2**, 312–332.

Himes, F. L., and S. A. Barber (1957), Chelating ability of soil organic matter, *Proc. Soil Soc. Am.* **21**, 368–373.

Huggins, M. L., and K. H. Sun (1946), Energy additivity in oxygen-containing crystals and glasses, *J. Phys. Chem.* **50**, 319–328.

Jackson, M. L., S. A. Tyler, A. L. Willis, G. A. Bourbeau, and R. P. Pennington (1948), Weathering sequence of clay-size minerals in soils and sediments. 1. Fundamental generalizations, *J. Phys. Coll. Chem.* **52**, 1237–1260.

Jackson, M. L., Y. Hseung, R. B. Corey, E. J. Evans, and R. C. Heuval (1952), Weathering sequence of clay size minerals in soils and sediments. 11. Chemical weathering of layer silicates. *Proc. Soil Sci. Soc. Am.* **16**, 3–6.

Jenny, H. (1941), *Factors in Soil Formation*. McGraw-Hill, New York.

Jenny, H. (1950), in *Origin of Soils. Applied Sedimentation* (P. D. Trask, ed.), pp. 41–61, Wiley, New York.

Jenny, H., and C. D. Leonard (1934), Functional relationships between soil properties and rainfall, *Soil Sci.* **38**, 363–381.

Keller, W. D. (1954), Bonding energies of some silicate minerals, *Am. Mineralogist* **39**, 783–793.

Keller, W. D. (1957), *The Principles of Chemical Weathering*. Lucas Bros., Columbia, Missouri.

Keller, W. D., W. D. Balgord, and A. L. Reesman (1963), Dissolved products of artificially pulverized silicate minerals and rocks, Pt. 1, *J. Sed. Petrol.* **33**, 191–204.

Keller, W. D., and A. F. Frederickson (1952), Role of plants and colloidal acids in the mechanism of weathering, *Am. J. Sci.* **250**, 594–603.

Krauskopf, K. B. (1959), The geochemistry of silica in sedimentary environments. Silica in Sediments, *Soc. Econ. Paleontologists Mineralogists Symp.* **7**, 4–18.

Lehman, D. S. (1963), Some principles of chelation chemistry, *Proc. Soil Soc. Am.* **27**, 167–170.

Livingstone, D. A. (1963), Data of geochemistry. Chapt. G. Chemical composition of rivers and lakes. *U.S. Geol. Surv. Profess. Paper* **440-G**.

Loughnan, F. C. (1962), Some considerations in the weathering of the silicate minerals, *J. Sed. Petrol.* **32**, 289–290.

Loughnan, F. C. (1960), Further remarks on the occurrence of palygorskite at Redbank Plains, Qld, *J. Roy. Soc. Queensland* **71**, 43–50.

Loughnan, F. C., and P. Bayliss (1961), The mineralogy of the bauxite deposits near Weipa, Queensland, *Am. Mineralogist* **46**, 209–217.

Loughnan, F. C. (1966), A comparative study of the Newcastle and Illawarra Coal Measure sediments of the Sydney Basin, N.S.W., *J. Sed. Petrol.* **36**, 1016–1025.

Marshall, C. E. (1964), *The Physical Chemistry and Mineralogy of Soils*. Vol. 1. *Soil Materials*. Wiley, New York.

Marshall, C. E., and C. A. Krinbull (1942), The clays as colloidal electrolytes, *J. Phys. Chem.* **46**, 1077.

Mason, B. (1952), *Principles of Geochemistry*. Wiley, New York.

McClelland, J. E. (1950), The effect of time, temperature and particle size on the release of bases from some common soil-forming minerals of different crystal structures, *Proc. Soil Soc. Am.* **15**, 301–307.

McConnell, D. (1951), Mechanisms of weathering, *Bull. Geol. Soc. Am.* **62**, 700.

McKenzie, R. C. (1963), Retention of exchangeable ions by montmorillonite, *Intern. Clay Conf.*, Pergamon Press, New York, pp. 183–193.

Merrill, G. P. (1906), *A Treatise on Rocks, Rock Weathering and Soils*. Macmillan, New York.

Pauling, L. (1940), *The Nature of the Chemical Bond*, 3rd ed. Cornell Univ. Press, Ithaca, New York.

Pedro, G. (1961), An experimental study on the geochemical weathering of crystalline rocks by water, *Clay Min. Bull.* **26**, 266–281.

Pettijohn, F. J. (1941), Persistence of heavy minerals and geologic age, *J. Geol.* **49**, 610–625.

Pierce, R. S. (1953), Oxidation and reduction potential and specific conductance of groundwater. Their influence on natural forest distribution, *Proc. Soil Sci. Soc. Am.* **17**, 61–67.

Polynov, B. B. (1937), *The Cycle of Weathering*. (A. Muir, transl.). Thos. Murby. London.

Radoslovich, E. W. (1960), The structure of muscovite, *Acta Cryst.* **13**, 919–932.

Reiche, P. (1943), Graphic representation of chemical weathering, *J. Sed. Petrol.* **13**, 58–68.

Schatz, A., V. Schatz and J. J. Martin (1957), Chelation as a biochemical weathering agent, *Bull. Geol. Soc. Am.* **68**, 1792–1793.

Sherman, G. D. (1952), The titanium content of Hawaiian soils and its significance, *Proc. Soil Sci. Soc. Am.* **16**, 15–18.

Stevens, R. E., and M. K. Carron (1948), Simple field test for distinguishing minerals by abrasion pH, *Am. Mineralogist* **33**, 31–49.

Tamm, O. (1930), Experimentelle Studien uber die Verwitterung und tonbildung von Feldspaten, *Chemie der Erde* **4**, 420–430.

Tiller, K. G. (1958), The geochemistry of basaltic materials and associated soils of southeastern South Australia, *J. Soil Sci.* **9**, 225–241.

van der Marel, H. W. (1954), Potassium fixation in Dutch soils. Mineralogical analyses, *Soil Sci.* **78**, 163–179.

Walker, P. (1960), A soil survey of the County of Cumberland, Sydney Region N.S.W., *Soil Surv. Unit New South Wales Bull.* **2**.

Wallace, A., R. T. Mueller, O. R. Lunt, R. T. Ascroft and L. Shannon (1955), Comparison of five chelating agents in soils, in nutrient solutions and in plant responses, *Soil Sci.* **80**, 101–108.

Wear, J. I., and J. L. White (1951), Potassium fixation in clay mineral studies as related to crystal structure, *Soil Sci.* **71**, 1–14.

White, D. E., J. D. Hem and G. A. Waring (1963), Data of Geochemistry. Chapt. F. Chemical Composition of subsurface waters, *U.S. Geol. Surv. Profess. Paper* **440-F.**

Williams, D. E., and N. T. Coleman (1950), Cation exchange properties of plant root surfaces, *Plant and Soil* **2**, 243–256.

Wright, J. R. and M. Schnitzer (1963), Metallo-organic interractions associated with podsolisation, *Proc. Soil Soc. Am.* **27**, 171–176.

IV. Environmental Factors Influencing Chemical Weathering

In the previous chapter it was established that chemical weathering proceeds through hydrolysis at the mineral surfaces followed by replacement of the metallic ions by hydrogen. Moreover, these reactions rapidly attain a state of equilibrium unless at least some of the released constituents, which may be either the metallic ions or polymerized silica units or both, are leached from the environment. In this chapter it is proposed to examine briefly the environmental factors which accelerate or retard leaching and hence influence the rate of chemical weathering.

THE ROLE OF CLIMATE

Climate is a paramount factor in chemical weathering. Rainfall controls the supply of moisture for chemical reactions and for the removal of soluble constituents of the minerals while temperature considerably influences the rate of these reactions. According to "van't Hoff's rule," for each 10°C rise in the temperature the velocity of a chemical reaction increases by a factor of from 2 to 3.

The influence of climate is manifest by a comparison of the weathering products of desert and of tropical rain forest areas. In arid regions where evaporation exceeds rainfall, water may penetrate the rocks but, during the ensuing long dry spell, it is returned to the surface and ultimately becomes lost through evaporation. As a result, soluble constituents of the rocks are not removed and reactions are slowed down accordingly. Moreover, not only is vegetation scant but, in addition, plant debris is quickly destroyed by the oxidizing atmosphere brought about by high temperatures and the considerable depth to the water table. Consequently, such areas are characterized by an adundance of unaltered or partly altered parent minerals, the presence of salts such as gypsum and carbonates, alkaline pH values (7.5–9.5), and a general paucity of organic matter. The oxidizing conditions ensure retention of the iron in the ferric state and this imparts red, brown, and yellow hues to the surfaces of rocks. The characteristic secondary products are montmorillonite, illite, and chlorite or, more probably, mixed layers of these minerals. Kaolinite is rare unless present in the parent rock, in which case it persists unaltered in composition. In the colder deserts, however, oxidation is retarded and a little organic matter may accumulate with the partly decomposed parent materials and the secondary products, producing gray rather than reddish colors.

In contrast, the rocks of the humid areas are generally well leached through the continual downward movement of the percolating waters, and the soluble products of the hydrolyzing reactions taking place at mineral surfaces are

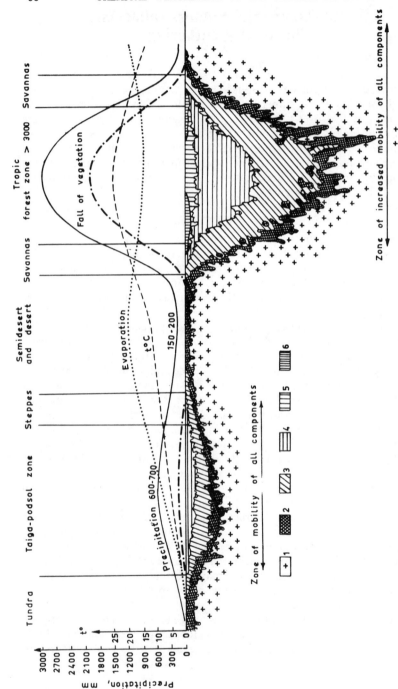

Figure 30.   Sketch of formation of weathering mantle in areas that are tectonically inactive (1) Fresh Rock; (2) zone of gruss eluvium, little altered chemically; (3) hydromica-montmorillonite-beidellite zone; (4) kaolinite zone; (5) ocher, Al_2O_3; (6) soil armor, $Fe_2O_3 +$ Al_2O_3. After Strakhov (1967); used by permission of Oliver & Boyd Ltd.

carried down to the water table and ultimately lost through subsurface drainage. Because of the ample supply of moisture, the water table is often at shallow depth and much of the altered zone may be permanently saturated and hence in a reduced state. Profuse vegetation may lead to extensive accumulations of organic matter above the water table, depending on the rate of supply and the degree of oxidation, and with accompanying leaching of the bases, low pH values (3.5–5.5) may result. Under these conditions chemical weathering tends to proceed rapidly through the loss of soluble constituents and the residue becomes progressively enriched in minerals containing a high proportion of alumina (kaolinite, halloysite, gibbsite, and boehmite) and titania (anatase and rutile). Ferric oxide minerals (hematite and goethite) may become dominant constituents if the rate of destruction of the organic matter is right.

Strakhov (1967) has given some interesting calculations of the effect of climate on the rate of weathering of silicate rocks. "The extent to which the decomposition of silicates and the removal of SiO_2 from the weathering substratum are slowed down is shown by the following computations (Ginsburg, 1958). A slab of fresh bedrock, one meter square and one centimeter thick, containing 47% SiO_2 would weigh 25,000 g or 25×10^6 mg. If the content of SiO_2 be reduced to 27% of the weight of this slab during the weathering process, 5.5×10^6 mg of SiO_2 would be lost. If the annual rainfall be 3750 mm, half of which percolates through the weathering slab and removes 10 mg/liter of SiO_2 by the water, 300 years would be required to remove the calculated quantity of silica from the slab. In a moist climate with a rainfall of 250–300 mm per year, 2200–4400 years would be required to remove the same quantity of silica, that is, 7–14 times the period required in the tropical zone. The actual retardation of the weathering process will be even greater, because the average temperature in temperate latitudes is 15°–20° lower than in the tropics; this means that the weathering and leaching in temperate latitudes takes place at one-twentieth to one-fortieth that in the moist tropics."

The intensity of weathering and the nature of the weathering products in relation to climate is well illustrated in Fig. 30. This diagram represents a composite of a number of weathering environments from nontectonic areas, along a north-south section stretching from the Arctic tundra to the forests and savannas of the tropics. The diagram also serves to illustrate another important aspect of weathering, namely the change in environmental conditions with depth. Thus in the tropic forest zone, the highly leached surface layers consist of a residue of ferric oxide and alumina (laterite and bauxite) minerals but, as leaching decreases at depth, sufficient silica is retained to combine with alumina forming kaolinite, while at still greater depth, potash and possibly both magnesia and ferrous iron persist in the structures of illites and montmorillonites. In brief, not one but a series of weathering environments exists superimposed on one another with the residual products of each forming the parent material for the succeeding environment.

However, it is not so much the total rainfall which governs the rate of weathering but rather that proportion of the total rainfall which actually infiltrates the weathering zone, percolates downward, and ultimately finds its way by subsurface drainage to creeks, rivers, lakes, or the ocean, carrying with it dissolved constituents. The proportion may vary widely from area to area depending on many factors. Thus, for example, it is most likely to be at a maximum where the rainfall is evenly distributed and occurs as gentle persistent showers, and at a minimum where the rainfall is seasonal and restricted to short violent downpours.

Temperature plays a dual role in chemical weathering. On the one hand, increasing temperatures promote weathering by greatly accelerating the chemical reactions. This is shown in Fig. 31, in which the clay content of

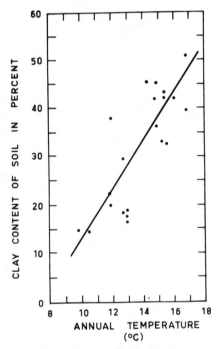

Figure 31. The clay content of soils derived from basic rocks in relation to mean annual temperatures (from H. Jenny, *Factors in Soil Formation* (1941); used by permission of McGraw-Hill Book Company).

soils derived from basic igneous rocks, mainly diorite and gabbro, extending along a north-south line from New Jersey to Georgia, is plotted against the mean annual temperature. The annual rainfall for all samples lies between 40 and 50 inches. The trend is accompanied by a change in the chemical composition of the soils, the $SiO_2 : Al_2O_3$ ratio being greater than 2.0 in the north and less than that value in the south. On the other hand, however, increasing temperatures create greater evaporation and hence tend to retard chemical weathering by reducing the quantity of water leaching through the weathering zone.

According to Crowther (1930), the influence of rainfall and temperature can be linked by what he termed the *leaching factor*, which is defined as follows:

$$\text{Leaching factor} = R - 3.3T$$

where R is the mean annual rainfall in centimeters and T the mean annual temperature in degrees centigrade. This is similar to the more widely accepted NS quotient of Meyer, which is given by the ratio of the mean annual rainfall in millimeters to the mean absolute saturation deficit of the air expressed in millimeters of mercury. Jenny (1941) considered the boundary between arid and humid climates to occur at an NS quotient of 200–250. However, it must be borne in mind that values obtained by these and other suggested empirical formulas are applicable on a broad zonal basis and no account is taken of the effects of variations in topography and parent material.

THE ROLE OF TOPOGRAPHY

Topography has a marked effect on the rate of chemical weathering and on the nature of the weathered products. It exerts this influence in several ways, by controlling (a) the rate of surface runoff of rain water and hence the rate of moisture intake by the parent rock, (b) the rate of subsurface drainage and therefore the rate of leaching of the soluble constituents, and (c) the rate of erosion of the weathered products and thereby the rate of exposure of fresh mineral surfaces.

On very steep slopes most of the rain water is lost through surface runoff and little penetrates the parent rock. At the same time erosion by running water, wind, landslides, and other agents is particularly active. Consequently, in such environments, mechanical disintegration of rocks proceeds at a much greater rate than chemical breakdown and only superficial accumulations of secondary products result.

Flat, low-lying areas, on the other hand, experience little surface runoff, and infiltration of rain water is at a maximum. However, in this type of environment, subsurface drainage tends to be sluggish and soluble products released by the hydrolyzing reactions persist in the environmental waters, thus inhibiting further breakdown of the parent minerals. Where the parent rocks are relatively rich in the alkalies or alkaline earths, or where the drainage waters from nearby hills are charged with such ions, the environment may become distinctly alkaline. The water table is at shallow depth and locally may rise above ground level forming swamps and promoting the accumulation of organic matter. A strongly reducing environment results.

The ideal conditions for chemical weathering are attained on rolling to gently sloping uplands where surface runoff is not excessive and the subsurface drainage is unimpeded. Under such conditions, the weathered zone may extend to a depth of 100 feet or more. However, even here local variations in topography create distinctly different environments which may find expression in contrasting weathered products.

The effective temperature of an area is also modified by the surface relief. For every 50 meters in elevation the temperature decreases by 1°C while steep slopes may considerably reduce the amount of sunlight reaching certain areas. The latter point is well illustrated by the comparative temperature measurements obtained by Pallman and Frei (Robinson, 1951) for two soils with contrasting aspects in a Swiss forest (Table 19).

TABLE 19
The Influence of Aspect on Soil Temperature[a]

Site			Equiv. temp. at 10 cm depth(°C)			
Height (meters)	Slope	Aspect	Winter	Spring	Summer	Autumn
1910	30°	S	3.2	12.6	17.4	17.7
1900	32°	NNE	−0.8	6.2	9.7	9.3

[a]After Pallman and Frei, quoted by Robinson (1951).

THE ROLE OF PARENT MATERIAL

Although parent material is not in itself an environmental factor, nevertheless variations in its texture, structure, and composition can exert a considerable influence on the rate of leaching.

Texture influences permeability and therefore the degree of infiltration of rainwater into the rock. Loose, friable sands, for example, are particularly permeable and the soluble constituents, such as carbonates, tend to be leached from areas above the water table. This is shown in Table 20, which represents a study by Salisbury (1925) of the sand dune system at Southport, Lancashire,

TABLE 20
Changes in pH and CaCO₃ Content of Dune Sands at Southport, Lancashire with Time[a]

Approx. age (years)	pH	CaCO₃ (%)
2	8.2	6.30
6	8.2	4.15
14	8.15	3.17
25	7.8	2.35
29	7.6	3.10
55	7.8	2.50
100	7.2	1.14
200	6.4	0.20
270	6.8	0.33
280	6.8	0.50
280	5.3	0.09

[a] After Salisbury (1925).

TABLE 21

Weathering Products in Relation to Some Specific Environments

Environment	pH	Eh	Behavior of metallic ions	Mineralogy
Nonleaching, hot. Rainfall 0–12 in.	Alkaline	Oxidizing	Some loss of alkalies. Iron present in ferric state.	Partly decomposed parent minerals. Illite, chlorite, montmorillonite, and mixed layered clay minerals. Hematite, carbonates and salts. Organic matter absent or sparse.
Nonleaching below the water table	Alkaline to neutral	Reducing	Some loss of alkalies. Iron present in ferrous state.	Partly decomposed parent minerals. Illite, chlorite, montmorillonite, and mixed layered clay minerals. Siderite and pyrite. Organic matter present.
Moderate leaching temperate climate. Rainfall 25–50 in.	Acid	Oxidizing to reducing	Loss of alkalies and alkaline earths. Some loss of silica. Concentration of alumina, ferric iron, and titania.	Kaolinite with or without degraded illite. Some hematite present. Organic matter generally present.
Intense leaching, hot. Rainfall > 50 in.	Acid	Oxidizing	Loss of alkalies, alkaline earths, and silica. Concentration of alumina, ferric oxide, and titania.	Hematite, goethite, gibbsite, and boehmite with some kaolinite. Organic matter absent or sparse.
Intense leaching, cool. Rainfall > 50 in.	Very acid	Reducing	Loss of alkalies, alkaline earths, and much of the iron. Alumina, silica, and titania retained.	Kaolinite, possibly with some gibbsite or degraded illite. Organic matter abundant.

England. The ages of the ridges were determined partly from the study of old maps and partly by examination of the annular ring development of the dwarf willow *Salix repens* which quickly followed consolidation of the dunes. The dense, compact nature of clays, on the other hand, tends to inhibit penetration of water and greatly increases the loss through surface runoff. Fractures and zones of weakness such as joints, faults, cleavage, and bedding planes offer easy access to water and greatly accelerate the leaching process in their near vicinity in addition to providing channels for the subsurface drainage. Readily soluble rocks, such as limestones, may give rise to solution channels, subterranean caverns, and sink holes which tend to divert much of the rain water into the parent rocks and considerably decrease surface runoff. Again, the rate of rain water infiltration progressively increases with the degree of alteration of the parent rock. Consider, for example, the alteration of a fresh lava flow to a lateritic crust. In the initial stages most of the rain water striking the exposed rock surfaces is rapidly lost through surface runoff or evaporation but with the development of a thick, cellular lateritic crust, virtually all rain water is soaked up by the weathering zone. Consequently, the rate of weathering may alter, not because of a change in climate or surface relief, but merely through the development of a mature weathered zone.

SUMMARY

By controlling the degree of leaching and the redox potential, the interplay of climate, topography, and parent material directly controls the rate of chemical weathering and the nature of the weathered products. In Table 21, the products of chemical weathering in relation to a number of specific weathering environments are summarized.

REFERENCES

Crowther, E. M. (1930), The relationship of climate and geological factors to the composition of the clay and the distribution of soil types, *Proc. Roy. Soc.* **107**, 10–30.

Jenny, H. (1941), *Factors in Soil Formation.* McGraw-Hill, New York.

Robinson, G. W. (1951), *Soils: Their Origin, Constitution and Classification*, 3rd ed. Thos. Murby and Co., London.

Salisbury, E. J. (1925), Note on the edaphic succession in some dune soils with special reference to the time factor, *J. Ecol.* **13**, 322–330.

Strakhov, N. M. (1967), *Principles of Lithogenesis*, Vol. 1 (J. P. Fitsimmons, S. I. Tomkeieff, and J. E. Hemingway, translators). Oliver and Boyd, London.

V. Chemical Weathering of Various Rock Types

INTRODUCTION

Advances in analytical techniques, particularly the more widespread application of X-ray diffraction, over the past two decades have given impetus to the study of the mineralogical transformations and the chemical reactions which accompany the weathering of minerals and rocks. Nevertheless, despite this renewed activity, our present state of knowledge still leaves much to be desired. Probably one of the contributing factors has been the tendency on the part of many workers to restrict their studies to the final alteration products and to neglect the intermediate stages in the transition from the parent rock. The alteration of basalt to a lateritic bauxite is not a single process in which the feldspar, pyroxene, and olivine of the basalt are pseudomorphed by gibbsite, hematite, goethite, and anatase, but rather several stages are involved. Generally the primary minerals are converted initially to clay minerals and then, with the ultimate loss of all potentially mobile constituents, the residue of oxides and hydroxides crystallizes out. To understand the mechanism of transformation of a parent rock such as basalt to a surface concentration of bauxite and laterite minerals, each and every stage must be examined. However, it should be borne in mind that the alteration products at a specific level in a weathered sequence were derived essentially from a parent rock at that level and that such a parent rock may have differed mineralogically, chemically, and texturally from the unaltered rock located below the present weathered sequence. To state this another way, the value of the data obtained from a chemical and mineralogical study of the alteration products at various levels within a weathered sequence depends on the original homogeneity of the rock. In general, homogeneity is realized most with the coarse-grained igneous rocks and least with thinly bedded, alternating sedimentary strata. However, even with the coarse-grained igneous rocks such as granites and gabbros, dykes of aplite and pegmatite may be common while deuteric and hydrothermal alteration can be such that the assumption of a homogeneous parent rock is not tenable.

Difficulties presented by variations in the composition of the parent rock may be partly alleviated by the examination of a number of weathered sequences from the one area and by the use of a satisfactory reference constituent. To be effective, the reference constituent should be present in both parent rock and altered products in sufficient quantities to minimize possible analytical errors and, in addition, should be completely immobile under the prevailing weathering conditions. As discussed in Chapter III, the only constituents which could possibly meet these requirements are ferric iron, titanium, and aluminum. Of these, ferric iron is unstable under reducing conditions and is converted to the potentially mobile ferrous state, while titanium rarely exceeds a few percent of either the parent rock or the weathered products, and moreover, if released from primary silicate minerals, may exhibit some degree of mobility. Aluminum, on the other hand, is immobile provided the environmental pH lies within the range 4.5–9.0 and is generally, but not invariably, a major constituent of both parent rock and weathered products. Consequently, aluminum would appear the most suitable reference constituent where the environmental waters are neither excessively acid nor alkaline.

In general, alteration is greatest at the surface and decreases in intensity downward toward the unaltered parent rock. This trend is attributed to the degree of saturation of the percolating waters. Rain water, on entering the surface, contains little of the potentially soluble constituents of the rock or soil but, with its passage downward, saturation with respect to these constituents increases and, consequently, their rate of solution decreases. In regions of high evaporation and sluggish subsurface drainage, much of the water may return by capillary action to the surface leaving behind concentrations of the soluble constituents at specific levels within the weathered zone. Moreover, plants assimilate mineral matter of the soil through their root systems and, upon decay of the plant and ultimate oxidation of the organic material, this mineral matter accumulates at the surface. The enrichment of potassium in the A horizon of many soils appears to have been accomplished by this mechanism. The translocation of mineral matter is probably assisted also by the activities of bacteria, worms, and other organisms.

BASIC CRYSTALLINE ROCKS

The basic crystalline rocks include the peridotites, gabbros, dolerites, basalts, and low-silica metamorphic rocks such as calcic hornfels. The dominant minerals are olivine, pyroxenes, and calcic plagioclases (labradorite to anorthite) with various accessories such as ilmenite, magnetite, and titanomagnetite.

Olivine is composed of discrete silica tetrahedra bonded together by magnesium and ferrous ions in octahedral coordination. Both octahedral cations are potentially mobile and their loss from the surface of the mineral causes the ready release of the individual tetrahedral units, thereby exposing fresh surfaces to attack. Consequently olivine decomposes rapidly. However, the rate of release of the silica may exceed its rate of solution, in which case the residual silica polymerizes into sheets and apparently "fixes" some of the magnesia, yielding serpentine as the crystalline phase. Generally, a complete balance of silica and magnesia is attained so that neither free silica nor free magnesia remains within the weathering zone. This and the ease with which released silica can polymerize into the sheet structure of the phyllosilicates are phenomena which may be observed repeatedly in the weathering of the silicate minerals. Much of the ferrous iron, if released above the water table, is oxidized, in which state it crystallizes or tends to crystallize as either the oxide, hematite and more rarely maghemite, or the hydroxide, goethite. However, much may persist in an extremely fine state and appear amorphous to X-radiation.

forsterite   serpehtine

$$2\ Mg_2SiO_4\ +\ 2H_2O\ \rightarrow\ Mg_3Si_2O_5(OH)_4\ +\ MgO$$

fayalite   hematite

$$2Fe_2SiO_4\ +\ O_2\ \rightarrow\ 2Fe_2O_3\ +\ 2SiO_2$$

As the weathering intensity increases, magnesia is preferentially leached from the serpentine which becomes progressively unstable and is converted to a new silica-enriched phase, saponite. It is not known whether complete destruction of the serpentine lattice is necessary for this transformation. Ultimately saponite becomes unstable and both the remaining magnesia and silica are leached, leaving a residue of oxides and hydroxides of iron.

Schellmann (1964) has described a complete weathering sequence developed on a serpentine-rich rock in Kalimantan, Borneo (Fig. 32 and Table

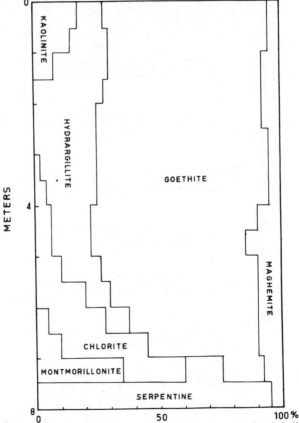

Figure 32. Mineralogy in relation to depth at Kalimantan, Borneo (after Schellmann, 1964).

TABLE 22

Chemical Analyses of the Weathering Sequence Developed on Serpentine at Kalimantan, Borneo[a]

Sample	Depth (meters)	Analysis (%) SiO$_2$	MgO	Fe$_2$O$_3$	Al$_2$O$_3$	Cr$_2$O$_3$	NiO	MnO	TiO$_2$	S	P$_2$O$_5$	Ignition loss
75	0.0–0.5	7.5	0.29	66.1	9.6	2.25	0.34	0.23	0.32	0.09	0.095	12.9
79	0.5–1.0	7.6	0.29	64.2	12.2	2.14	0.39	0.37	0.35	0.13	0.095	12.5
80	1.0–1.5	4.61	0.43	64.8	13.1	1.81	0.60	0.77	0.29	0.20	0.059	13.3
81	1.5–2.0	2.86	0.44	65.9	14.6	1.85	0.85	0.85	0.26	0.29	0.018	13.5
82	2.0–2.5	2.61	0.60	67.6	13.0	2.04	0.85	0.95	0.23	0.29	0.011	13.1
83	2.5–3.0	2.22	0.70	68.2	13.1	2.13	0.92	0.96	0.20	0.30	0.011	13.1
84	3.0–3.5	2.28	0.89	68.2	13.3	2.29	0.95	0.91	0.20	0.26	0.101	12.6
85	3.5–4.0	2.83	0.75	68.1	12.4	2.43	0.87	0.08	0.19	0.26	0.204	12.9
86	4.0–4.5	2.79	1.23	70.2	10.6	2.61	0.96	0.93	0.16	0.24	0.007	12.2
87	4.5–5.0	2.38	1.23	69.7	10.2	3.00	1.23	1.23	0.14	0.21	0.003	11.4
88	5.0–5.5	4.95	3.87	64.9	10.8	2.88	1.16	0.99	0.15	0.16	0.003	11.9
89	5.5–6.0	7.8	8.16	59.1	9.5	2.67	0.94	0.91	0.15	0.11	0.005	11.1
90	6.0–6.5	9.0	7.06	57.2	9.1	2.54	0.82	0.93	0.15	0.08	0.004	11.4
91	6.5–7.0	19.0	7.58	50.0	8.4	2.13	1.18	0.73	0.14	0.07	0.001	10.6
92	7.0–7.5	34.0	11.4	34.9	5.6	1.41	1.81	0.54	0.10	0.04	0.003	9.7
KG14	7.5	39.4	33.2	8.4	2.5	0.27	2.10	0.13	0.02	—	0.001	13.3

[a] After Schellman (1964).

22). In the early stages, the serpentine alters to montmorillonite (saponite) and chlorite but, as leaching intensifies, the montmorillonite and, later, the chlorite are destroyed. Alumina, present to the extent of 2.5% in the unaltered serpentine, is progressively concentrated in the weathering zone permitting the early development of chlorite, but upon the loss of magnesia and silica from this mineral, gibbsite forms. Apparently, near the surface resilification of some of the gibbsite occurs yielding kaolinite. A very marked concentration of ferric minerals, goethite, and maghemite appears toward the base of the weathering zone and increases upward, reaching a maximum between 4.0 and 4.5 meters, and then declines slightly toward the surface Schellmann considered enrichment at the 4.0–4.5 meter level to be due to transport of iron from above by descending waters. The plot of the chemical data against depth in Fig. 33 reflects the mineralogy shown in Fig. 32. The

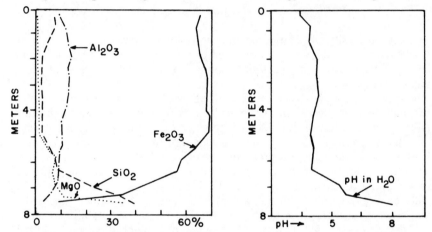

Figure 33.   Variations in oxide contents and pH with depth at Kalimantan, Borneo (after Schellmann, 1964).

pH is distinctly alkaline at the base of the section, but with the development of saponite it becomes mildly acid, attaining a minimum of 5.0 in the highly leached surface zone, probably owing to the presence of H^+-kaolinite.

Craig (1963) has shown that olivine may alter directly to magnesium-rich montmorillonite without passing through the serpentine stage, while Sherman *et al.* (1962) found nontronite the initial product of the weathering of olivine at Lualualei, Hawaii.

In the pyroxenes, the tetrahedra are arranged in chains which are bonded together by metallic ions, the most common being those which enter into octahedral coordination with oxygen, such as Mg^{++}, Fe^{++}, Fe^{3+}, and Al^{3+}. Some aluminum may replace silicon in the tetrahedra. Bonding by the octahedral cations is relatively weak and pronounced cleavages with planes approximately normal to each other parallel the silica chains. Access of water by way of the cleavages promotes solution of the bonding cations and causes rapid breakdown of the structure. Upon release, the chains tend to polymerize into sheets incorporating residual alumina and magnesia, forming chlorite or montmorillonite or both. Ferrous iron is readily stabilized by oxidation to the ferric state, and at the same time titania crystallizes as anatase. Where the release of calcium from breakdown of the pyroxene exceeds its rate of solution, calcite also develops. A general weathering sequence for the pyroxenes is shown in Fig. 34.

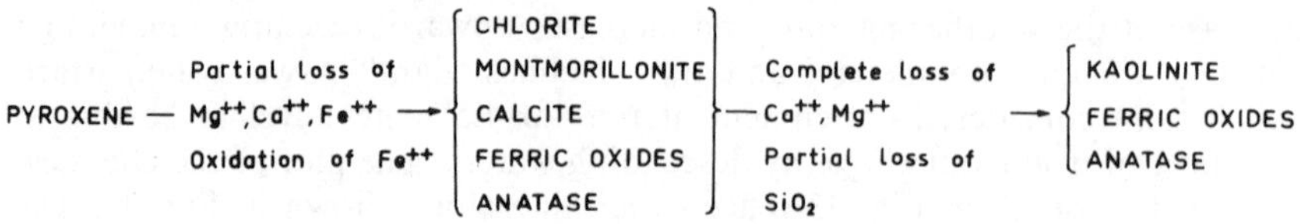

$$
\text{PYROXENE} - \begin{array}{c} \text{Partial loss of } Mg^{++}, Ca^{++}, Fe^{++} \\ \text{Oxidation of } Fe^{++} \end{array} \longrightarrow \left\{ \begin{array}{l} \text{CHLORITE} \\ \text{MONTMORILLONITE} \\ \text{CALCITE} \\ \text{FERRIC OXIDES} \\ \text{ANATASE} \end{array} \right\} \begin{array}{c} \text{Complete loss of } Ca^{++}, Mg^{++} \\ \text{Partial loss of } SiO_2 \end{array} \longrightarrow \left\{ \begin{array}{l} \text{KAOLINITE} \\ \text{FERRIC OXIDES} \\ \text{ANATASE} \end{array} \right.
$$

Figure 34.   Weathering sequence for some common pyroxenes.

As the intensity of weathering increases, all the lime and magnesia are lost together with the silica not required to saturate alumina, and the residue becomes progressively enriched in kaolinite, anatase, and oxides and hydroxides of ferric iron.

In the framework structure of the feldspars, the tetrahedra are linked through all four oxygen atoms to yield a three-dimensional configuration. Aluminum substitution for silicon within the tetrahedra creates a net negative charge on the framework which is balanced by an alkali or calcium ion. Breakdown of the feldspars proceeds through the loss of these alkali or calcium ions, but for their escape, the tetrahedral framework must be raptured. In the alkali feldspars the aluminum-to-silicon ratio is 1:3, and in the calcic feldspars, 1:1. Despite the similarities in structure of the alkali and calcic feldspars, a considerable disparity exists in their weathering stability as shown by Goldich (1938). Apparently, the greater replacement of silicon by aluminum in anorthite results in a higher degree of polarization of the oxygens and the lattice is more prone to rupture at the Al–O–Si bonds. With the loss of the metallic ions, the framework structure breaks down into chains which tend to polymerize into sheets in much the same manner as those released from the pyroxenes. Where the leaching conditions are inadequate to remove magnesium and ferrous ions as rapidly as they are released from the breakdown of associated minerals, these ions tend to be "fixed" by the new residual structure and montmorillonite or chlorite or both results. Chapman and Greenfield (1949), Smith (1962), and Craig and Loughnan (1964) have described the alteration of calcic feldspars to montmorillonite.

Magnetite, titanomagnetite, and ilmenite are the more common accessory minerals of the basic crystalline rocks. These minerals oxidize readily during weathering to yield hematite, maghemite, and (or) goethite, simultaneously releasing titania, which may have limited mobility as observed by Sherman (1952a) and Craig and Loughnan (1964), but ultimately crystallizes as anatase, often in the polycrystalline form known as leucoxene. In a study of the weathering of fine-grained dolerite dykes in the Sydney area of Australia, Loughnan and Golding (1957) found unusual skeletal octahedra of leucoxenic anatase in the residual clay, which consisted of well-crystallized kaolinite with or without illite. The leucoxenic octahedra ranged from 0.1 to 0.2 mm in size (Fig. 35) and were believed to have formed as pseudomorphs after titaniferous magnetite.

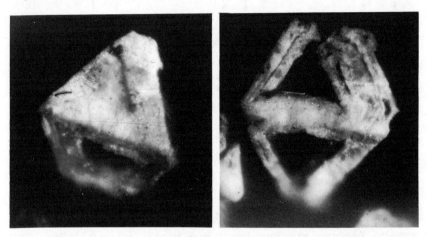

Figure 35.   Octahedral leucoxene pseudomorphs in clays derived from the weathering of dolerites, Sydney, Australia. (after Loughnan and Golding, 1957). × 300.

Weathering of Basalts in the Hawaiian Islands

The volcanic rocks of the Hawaiian Islands range in composition from trachytes with a silica content of 62% to alkali-enriched, silica-deficient basanites. However, the commonest type is a basic, olivine basalt, known as oceanite, in which the silica content approximates 45%. Under the prevailing tropical climate with a rainfall as high as 400 inches per annum, these rocks decompose rapidly.

Bates (1962) made a study of the weathering of these volcanic rocks and concluded that the controlling factors are age, rainfall, composition, and texture of the parent rock, and the topography. He found where leaching is moderate only, the plagioclase of the basalt alters to halloysite whereas the volcanic glass yields allophane, an amorphous hydrated aluminum silicate, together with alumina and silica gels. Simultaneously, olivine is converted to serpentine or a trioctahedral montmorillonite and various ferric hydrates and gels. However, where leaching approaches the optimum, gibbsite, laterite minerals, and amorphous alumina and ferric hydrates are the final products. These trends are shown in Fig. 36. Bates concluded that "Gibbsite is produced by (1) removal of silica from halloysite, (2) dehydration of Al-gel, and (3) precipitation from solution. Although it is possible that the mineral may form directly from feldspar, halloysite is the common crystalline

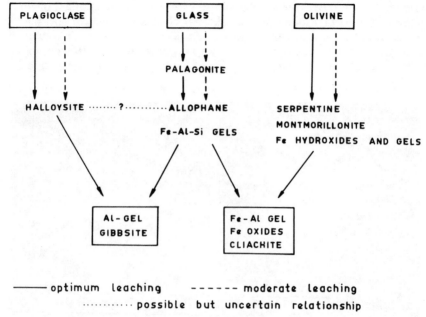

Figure 36.   Weathering sequence of basaltic lavas, Hawaiian Islands (after Bates, 1962).

intermediate on both the megascopic and microscopic scales. An amorphous transition state, probably ranging in composition from allophane to Al-gel, exists as part of the change from halloysite to gibbsite as evidenced by electron microscope and diffraction work on pseudomorphs after halloysite tubes found in certain samples studied in more detail than others.''

According to Sherman (1949, 1952b), two distinctive trends can be observed in the weathering of these basalts and the controlling factor is not so much the total rainfall but rather its distribution. Where a definite dry period exists, the end products of the alteration are considerably enriched in ferric oxide and titania, whereas in the absence of a dry period, alumina is the dominant constituent of the surface horizon. Apparently, in a continuously wet climate stabilization of the iron by oxidation and of the titania by desiccation of $Ti(OH)_4$ does not occur and appreciable quantities of both are lost through leaching. The influence of the rainfall distribution on the mineralogy of the weathered products is shown by a comparison of Figs. 38 and 39. Weathering in a wet and dry climate leads to the development of a titaniferous crust (Table 23) whereas in continuously wet areas the end product is dominated by bauxite minerals.

Lateritic Bauxites Developed from Basalt, County Antrim, Northern Ireland
    The occurrence of fossil lateritic bauxites intercalated with tertiary basalts in County Antrim, Northern Ireland has been described by Eyles (1952).

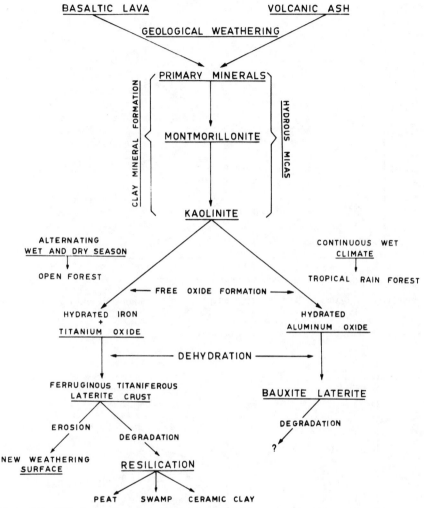

Figure 37.   Weathering sequence of basaltic lavas in the Hawaiian Islands (after Sherman, 1952b).

The bauxites apparently developed from the basalt during a prolonged interflow period by two distinct processes, (a) a preliminary stage in which magnesia, lime, and the alkalies were leached and the rock was converted to an assemblage of kaolin minerals, principally halloysite, and (b) a later stage of desilication whereby the kaolin minerals were replaced by a residue enriched in gibbsite. Chemical and mineralogical analyses of successive phases in these alteration processes are given in Table 24.

In the zone of greatest weathering, iron oxide tended to become segregated from alumina and an upper crust of highly ferruginous laterites, containing

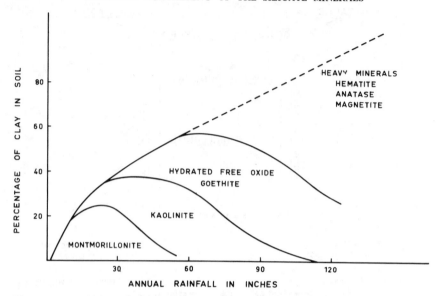

Figure 38. Progressive types of clay development in Hawaiian soils under an alternating
wet and dry climate (after Sherman, 1952b).

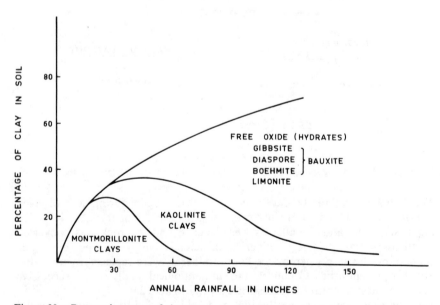

Figure 39. Progressive types of clay development in Hawaiian soils under a continuously
wet climate (after Sherman, 1952b).

TABLE 23

Chemical Composition of Typical Titaniferous Lateritic Crusts of Alternating Wet and Dry Areas in the Hawaiian Islands[a]

Mineralogy (%)

Location	pH	SiO_2	Al_2O_3	Fe_2O_3	TiO_2
Waimea Canyon, Kauai	4.2	4.64	13.65	48.50	24.85
Haiku, Maui	4.2	14.04	13.20	48.08	18.92
Lihue, Kauai	4.5	28.50	25.80	10.94	24.94
Pali, Molokai	5.0	12.60	12.60	52.04	14.04
Kehaha, Kauai	4.4	12.52	7.81	58.02	14.44

[a] After Sherman (1952a).

TABLE 24

Analyses of Successive Phases in the Alteration of Basalts in County Antrim, Northern Ireland[a]

	Analysis[b] (%)				
	1	2	3[c]	4[c]	5[c]
SiO_2	45.84	32.62	23.25	6.88	0.84
Al_2O_3	16.92	27.64	31.96	39.76	44.07
Fe_2O_3	3.73	19.88	24.55	26.65 }	} 27.12
FeO	5.85	0.90	0.71	—	
TiO_2	2.78	2.69	2.91	3.60	3.53
MgO	6.34	0.08	0.38 }	0.08	—
CaO	7.36	0.16	Trace }		—
Na_2O	1.96	Trace	Trace	—	—
K_2O	0.31	Trace	0.20	—	—
$H_2O + 105°C$	3.72	12.24	15.85	21.36	24.07
$H_2O - 105°C$	4.15	3.53	—		
TiO_2/Al_2O_3	0.16	0.10	0.09	0.09	0.08
Kaolinite	—	< 70	50.0	14.8	1.8
Gibbsite	—	—	18.7	51.9	66.4

[a] After Eyles (1952). Published by permission of Her Britannic Majesty's Stationery Office.
[b] (1) Parent basalt; (2) kaolinitic zone; (3) incipient lateritization; (4) advanced lateritization; (5) lateritic bauxite.
[c] Determined on a dry basis.

a high proportion of titania, formed above the bauxite horizon. This crust is similar to that described by Harder (1952) from the weathering of basalts in central India (Fig. 40).

Weathering of Basalts in New South Wales

Craig and Loughnan (1964) have described the chemical and mineralogical transformations accompanying the weathering of volcanic rocks in New

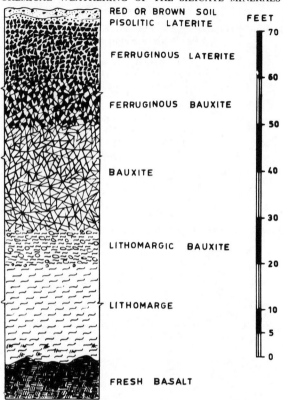

Figure 40.   Vertical section of a weathered sequence developed on basalt in central India
(after Harder, 1952).

South Wales ranging from shoshonite (trachyandesite) to olivine basalt in composition. From the detailed study of six weathered sequences, they concluded that olivine is the least stable of the primary silicates present in the basalts and that the pyroxenes, augite and titanaugite, are generally destroyed before the plagioclases. Sanidine, present in one rock only, proved remarkably stable to chemical weathering and persisted throughout the zone of weathering.

Olivine and the pyroxenes alter directly to trioctahedral montmorillonites, presumably of the saponite type, whereas a dioctahedral, aluminous montmorillonite is the initial alteration product of the plagioclases. In the near surface horizons, these montmorillonites are rendered unstable by the more intense leaching conditions and three degradation products are evident, namely halloysite, kaolinite, and poorly crystallized montmorillonite. Simultaneously, iron oxide as goethite, hematite, or an amorphous precursor of these is concentrated in the oxidized parts of the weathering zone. These weathering trends are shown diagrammatically in Fig. 41. In Fig. 42, the

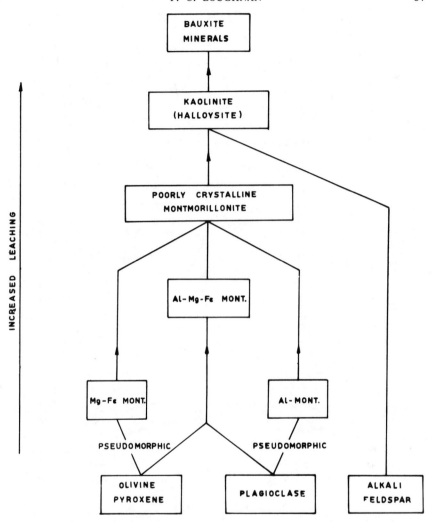

Figure 41.   Diagrammatic representation of the weathering sequence developed on Tertiary
basalts in New South Wales (after Craig and Loughnan, 1964).

mineralogy in relation to depth is given for one of these weathered sequences
from the Bathurst area where the average annual rainfall is 22 inches and the
mean annual temperature 57°F. The parent rock is an olivine basalt with an
intergranular to subophitic texture, and is composed of approximately 30%
plagioclase, 20% titanaugite, 25% olivine, 5% magnetite, and 20% glass.
Montmorillonites form the initial breakdown product, but higher in the section where leaching is more efficient these are replaced by kaolinite and halloysite. The ferruginous minerals increase up to the 3-foot level and then show a

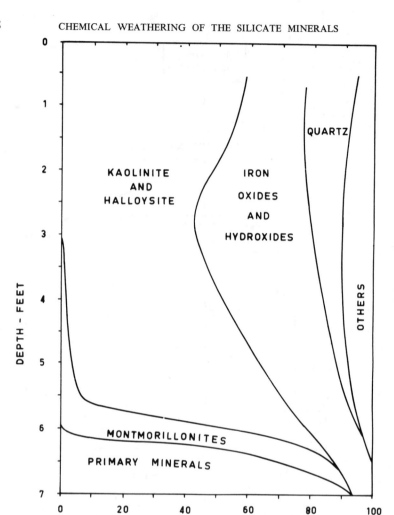

Figure 42.   Mineralogy in relation to depth at Bathurst, New South Wales (after Craig and Loughnan, 1964).

slight decline. Quartz appears at the 6-foot level and slowly increases toward the surface. The source of the quartz is not known.

The chemical data for the same weathered sequence are given in Table 25. In Fig. 43, these data have been recalculated on an anhydrous basis and the concentration ratio (i.e., the ratio of the content of an oxide at a specific level in the section to its content in the parent material) for each constituent has been plotted against depth from the surface. For convenience, alumina has been used as the reference constituent and given unit concentration ratio

TABLE 25

Chemical Data for the Weathered Sequence Developed on Basalt at Bathurst, New South Wales[a]

No.	Depth	Analysis (%)									
		SiO_2	Al_2O_3	Fe_2O_3	CaO	MgO	Na_2O	K_2O	TiO_2	H_2O	Total
B1	0ft 6in	39.3	20.3	16.4	1.7	0.5	0.3	0.6	2.4	18.6	100.1
B2	1ft 6in	36.1	20.6	21.8	0.5	0.3	0.2	0.2	4.1	16.3	100.1
B3	2ft 6in	27.8	18.5	33.2	0.5	0.2	0.2	0.1	5.2	14.3	100.0
B4	3ft 6in	28.9	19.4	30.5	0.5	0.2	0.2	0.1	4.8	15.8	100.4
B5	4ft 6in	33.4	21.4	24.3	0.5	0.3	0.2	0.1	4.1	15.8	100.1
B6	5ft 6in	37.1	23.2	18.1	0.6	0.3	0.2	0.1	3.8	16.8	100.2
B7	6ft 6in	42.2	15.2	14.0	9.0	8.7	1.1	0.9	2.0	7.2	100.3
B8	7ft 0in	45.0	14.6	13.3	9.8	10.1	2.4	1.5	1.9	1.6	100.2

[a]After Craig and Loughnan (1964).

throughout, the values for the other oxides being adjusted accordingly. It will be observed that, in agreement with the mineralogical data in Fig. 42, most of the soda, potash, lime, and magnesia and nearly half the silica have been lost between the 5- and 6-foot levels, whereas ferric oxide and titania show a marked concentration up to the 3-foot level and then diminish toward unity at the surface.

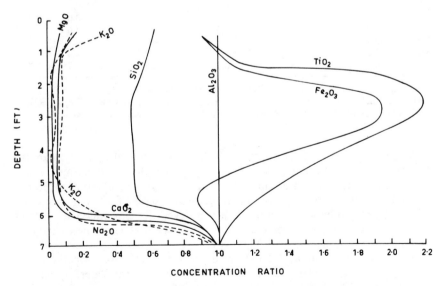

Figure 43. Concentration ratio versus depth for the weathered sequence developed on basalt at Bathurst, New South Wales (after Craig and Loughnan, 1964).

In the Sydney area, sandstones and shales of Triassic age have been intruded by numerous fine-grained dolerite dykes. Where the enclosing rock is shale of the Wianamatta Group, the dolerite is essentially unaltered but where the more permeable sandstones of the Hawkesbury form the host, the dolerites are converted to clay down to the level of the water table, which in places is at considerable depth. Loughnan and Golding (1957) determined the mineralogy of some of these dyke clays which had been exploited commercially in the past mainly for use in the local ceramic industry, and showed that several clay mineral assemblages occur, namely (1) kaolinite, (2) kaolinite and well-crystallized illite, and (3) mixed-layered illite-montmorillonite with or without discrete kaolinite. Generally the clays are characterized by a high titania content, often in excess of 5% (Table 26 and Fig. 35). However, in one of

TABLE 26

Chemical Analyses (in Percent) of Dyke Rocks and Derived Clays from the Sydney Area of Australia

| | Dolerite[a] | Clay[b] | | | |
		1	2	3	4
SiO_2	42.5	43.6	49.9	43.1	49.2
Al_2O_3	15.7	37.8	29.0	34.0	27.7
Fe_2O_3	3.5	0.3	2.3	0.7	0.9
TiO_2	1.9	3.0	5.9	5.7	0.5
Cr_2O_3	—	—	—	—	4.1
MgO	7.2	—	—	—	0.2
CaO	9.5	—	—	—	0.0
K_2O	1.8	2.7	2.3	0.7	4.6
Na_2O	3.1	—	—	—	0.1
H_2O+	2.2	9.6	9.5	12.3	7.4
H_2O-	0.6	2.6	0.3	3.0	4.9

[a] Average of 7 analyses of dolerite. After White and Mingaye (1904).
[b] (1, 2, 3.) Clays derived from weathering of dolerite. After Loughnan and Golding (1957).
(4) Chromium-bearing clay derived from weathering of dolerite. After Loughnan and Bayliss (1963).

these dyke clays from Cowan (No. 4, Table 26) examined by Loughnan and Bayliss (1963), the titania content is relatively low whereas the clay fraction composed of discrete illite and a random interlayering of montmorillonite and illite, is bright apple green and contains in excess of 4% chromic oxide within the clay mineral structures. They believed the clay was derived from an ultrabasic differentiate of the dolerite magma that contained chromian diopside rather than the usual titanaugite.

Bayliss and Loughnan (1963) have described two unusual weathered sequences developed from basalt in northern New South Wales, apparently during the middle Tertiary. In one of these, exposed in a shaft 65 feet deep near Emmaville (Fig. 44), the lowermost 4 feet (Zone 1) is composed of

MINERALOGY CUMULATIVE %

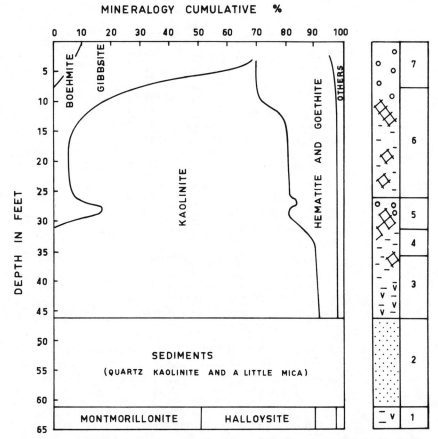

Figure 44. Mineralogy in relation to depth at Emmaville, New South Wales (after Bayliss and Loughnan, 1963).

montmorillonitic and halloysitic clay which still retains the vesicular structure of the parent basalt. The clay is overlain by 15 feet of kaolinitic sand of sedimentary origin (Zone 2) and this is succeeded by a further 11 feet of clay (Zone 3) which also possesses the residual structure of the parent basalt but, unlike that at the base of the shaft, consists essentially of kaolinite. Zone 3 grades imperceptibly upward into a 4-foot light-colored, mottled zone (Zone 4) and this, in turn, is overlain by 5 feet of yellow, bauxite clay (Zone 5). A well-defined boundary at the 26-foot level separates the top of Zone 5 from 18 feet of light red, mottled clay (Zone 6) which grades upward into a deep red pisolitic zone rich in gibbsite, hematite, and (or) goethite, approximately 8 feet thick (Zone 7).

In the other exposure, near Inverell (Fig. 45), kaolinite clay with a spheroidal structure characteristic of weathered basalt comprises the lowermost 8 feet.

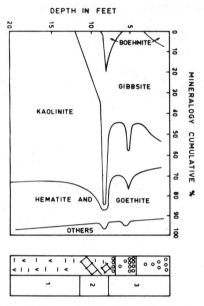

Figure 45.   Mineralogy in relation to depth at Inverell, New South Wales (after Bayliss and Loughnan, 1963).

This is succeeded by a clearly defined mottled zone, 4 feet thick, and 8 feet of deep red, pisolitic bauxite. Two persistent nodular bands, rich in gibbsite occur within the pisolitic zone at depths of 4.5 to 5.5 and 7.5 to 8.0 feet from the surface.

A possible explanation for the origin of these unusual features is that alteration of the individual flows occurred soon after emplacement and prior to the succeeding outpouring of basalt.

Weathering of Pyroclastics in the West Indies

In an investigation of the chemical weathering of pyroclastic materials, including andesitic glass, calcic plagioclases, and ferromagnesian minerals, on the island of St. Vincent in the West Indies, Hay (1959) found fully hydrated halloysite the principal alteration product. Volcanic glass proved the least stable constituent and where the deposits are only slightly weathered, it is the only constituent to show alteration. A small proportion of montmorillonite generally accompanies the halloysite in the alteration products. The anorthite content apparently influenced the rate of breakdown of the plagioclases, plagioclase An_{50-65} being considerably more stable than plagioclase An_{80-100}. Of the ferromagnesian minerals, olivine with a forsterite content of 74% decomposed more readily than augite, hypersthene, or olivine with 78% forsterite, while hornblende displayed about the same stability as plagioclase An_{50-65}. These findings are summarized in Table 27.

TABLE 27

Sequence of Weathering on St. Vincent Island[a,b]

	Crystal ash		Phenocrysts	
	Ferromagnesian minerals	Plagioclase	Ferromagnesian minerals	Plagioclase
Sequence of Alteration ↓	Fo_{74-78} (some)	An_{80-100}	Fo_{74}	An_{80-100}
	Fo_{74-78} (most)	An_{65-80}	Fo_{78}, Aug, Hyp	An_{65-80}
	Aug, Hyp			An_{50-65} (some)
	Hornblende	An_{50-65}	Hornblende	An_{50-65}

[a] After Hay (1959); used by permission of The University of Chicago Press.
[b] Fo=forsterite; An=anorthite; Aug=augite; Hyp=hypersthene.

ACID CRYSTALLINE ROCKS

The acid crystalline rocks, which include the granites, granite gneisses, porphyries, rhyolites, and quartz-feldspathic hornfels, are characterized by an abundance of potash- and soda-rich feldspars (albite to andesine) with quartz and lesser amounts of amphiboles and micas.

The potash- and soda-rich feldspars rank among the more stable of the crystalline silicates and require more intense leaching conditions or, alternatively, a greater interval of time for their breakdown than do the majority of other silicate minerals. Kaolinite and halloysite are the common weathered products although illite or montmorillonite or both may result where the environment is only mildly leaching (McKenzie, Walker, and Hart; 1949). Under particularly aggressive leaching conditions the alkalies and silica may be removed and the feldspars are pseudomorphed by bauxite minerals (Grant, 1963).

In a study of the genesis of residual kaolins derived from pegmatites in the southern Appalachians, Sand (1956) found that whereas both potash and plagioclase feldspar may alter to halloysite, frequently such alteration is restricted to plagioclase, the potash feldspars yielding initially illite and later kaolinite as weathered products. These reactions may be represented as follows:

$$\overset{\text{albite}}{2NaAlSi_3O_8} + 9H_2O \rightarrow \overset{\text{halloysite}}{Al_2Si_2O_5(OH)_4 2H_2O} + 2H_2SiO_3 + 2NaOH$$

$$\overset{\text{sanidine}}{3KAlSi_3O_8} + 8H_2O \rightarrow \overset{\text{illite}}{KAl_2(AlSi_3)O_{10}(OH)_2} + 6H_2SiO_3 + 2KOH$$

$$\overset{\text{illite}}{2KAl_2(AlSi_3)O_{10}(OH)_2} + 5H_2O \rightarrow \overset{\text{kaolinite}}{3Al_2Si_2O_5(OH)_4} + 2KOH$$

However, examination of the stability curve of halloysite by Bates (1952) reveals that the mineral forms only under water or in an environment of very high humidity and if exposed for any period in an evaporating atmosphere it tends to dehydrate irreversibly to metahalloysite (Fig. 46), in which form it is difficult to distinguish from b-axis disordered kaolinite. Probably much of

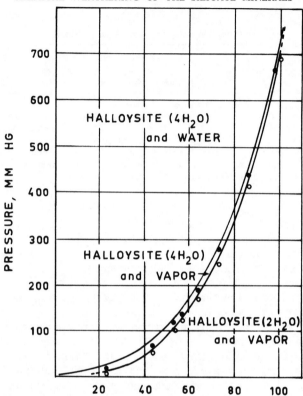

Figure 46. Phase relationships of halloysite (after Bates, 1952). Upper curve represents the vaporization of water; lower curve, the boundary between halloysite(4H₂O) and halloysite(2H₂O).

the kaolinite recorded as a product of the weathering of feldspars is in reality metahalloysite which has formed from halloysite during a period of desiccation.

In accounting for the mechanism of formation of illite from feldspars, De Vore (1958) has pointed out that the release of tetrahedral chains having a Si-Al ordering which could yield mica structures upon polymerization requires the severing of the least number of bonds in the feldspar structure. He has suggested that in the presence of a restricted supply of water, these chains have at least temporary stability and, if polymerization took place in the presence of potassium and octahedral cations such as aluminum, magnesium, ferric and ferrous iron, micas, and montmorillonites could form. However, if the feldspars were reduced to isolated tetrahedra, he considered the formation of these minerals unlikely, their place being taken by kaolins. The nature of the residual deposits from the weathering of feldspar, therefore, depends on the relative rates of breakdown of the mineral and removal of the released constituents. If the rate of breakdown exceeds the rate of removal, a residue of illite or montmorillonite, or both, can be anticipated; if the converse, kaolinite or halloysite, or both, is the more likely product.

The direct alteration of primary minerals to gibbsite without passing through a clay mineral phase may occur particularly where silica-undersaturated, alumina-enriched rocks, such as nepheline syenite, form the parent material. The high alkali-to-silica ratio of these rocks favors the preferential loss of silica and facilitates the development of gibbsite pseudomorphs after feldspar. Stephen (1962) has described the weathering of syenites at Mt. Zomba, Nyasaland, in which the feldspars, orthoclase, and plagioclase ($Ab_{90}An_{10}$) are pseudomorphed by gibbsite. From the available evidence he concluded that kaolinization succeeds rather than precedes, bauxitization.

Quartz is generally a prominent constituent in the acid crystalline rocks. It has a close-packed tetrahedral framework structure analogous to the feldspars, but unlike the latter it lacks readily mobile alkali or calcium ions and, in consequence, it is one of the more resistant of the silicate minerals to leaching solutions. It does not alter to form new crystalline phases but rather is slowly taken into solution. According to Siever (1957), the thermochemical data indicate that the solubility should lie between 7 and 14 ppm at ambient temperatures. This is approximately one-tenth the solubility of amorphous silica given by Krauskopf (1959), who has pointed out that a solution which is saturated with respect to amorphous silica is supersaturated with respect to crystalline silica and hence quartz should precipitate. However, a metastable equilibrium is established and a considerable period of time is required for precipitation. Although quartz has a low solubility at or near neutrality, the value nevertheless is measurably greater than that of alumina, ferric oxide, and titania, and under prolonged leaching conditions some loss of quartz relative to alumina can be anticipated.

The amphiboles, of which hornblende is the most common in the acid crystalline rocks, have a structure similar to the pyroxenes but differ in that hydroxyl groups are present and the tetrahedra are linked into double chains. The metallic ions bond to the hydroxyl groups as well as free oxygen of the tetrahedral chains. Two sets of cleavages parallel the lengths of the chains and weathering of the amphiboles generally proceeds initially by leaching along these fractures. According to Goldich (1938), hornblende is a little more resistant to breakdown than augite.

In a study of the weathering of hornblende in the Malvern Hills region of Great Britain, Stephen (1952) found that the mineral alters from the cleavage planes into the body of the crystal, yielding chlorite with some sphene, hematite, and epidote as the initial products. At this stage apparently all the silica and alumina and much of the lime, magnesia, and iron are retained in the secondary products. Further leaching results in the conversion of the chlorite first to a chlorite-vermiculite mixed layer, and finally to vermiculite. However, the work of Goldich (1938) on the weathering of an amphibolite from the Black Hills area of South Dakota indicates that hornblende may decompose to beidellite (aluminous montmorillonite) or related clay minerals through the loss of calcium. Alexander *et al.* (1941), on the other hand, believed the weathering of hornblende in amphibolites from North Carolina to be partly responsible for the development of gibbsite in that area. On this basis, therefore, it would appear that the amphiboles have much the same weathering sequence as the pyroxenes (Fig. 34).

In the common micas, aluminum in fourfold coordination replaces silicon to the extent of 1:4 and the tetrahedra are polymerized into sheets. Two sheets bonded by cations in octahedral coordination comprise a single layer. Because of the aluminum replacement of silicon, the layers have a net negative charge of one valency unit and this is balanced by a potassium ion located between the layers. The potassium-oxygen bond is weak and a pronounced cleavage develops parallel to the layers. Weathering proceeds initially along these zones of weakness by solution of the potassium and the introduction of water molecules, a process which enables the potassium-depleted structure to expand in a manner similar to montmorillonite.

A considerable disparity exists in the rate of solution of potassium ions from the dioctahedral (muscovite) and trioctahedral (biotite-phlogopite series) micas. This has been attributed to (a) differences in the orientation of the O–H bond of the hydroxyl groups, (b) the ready oxidation of ferrous ions in the octahedral sheet to the ferric state, thus upsetting the balance of charges within the structure, and (c) the preferential solution of octahedral ferrous and magnesium ions from the trioctahedral micas.

According to Bassett (1960) the greater susceptibility of the trioctahedral micas to alteration relative to muscovite is attributable to the orientation of the hydroxyl groups of the mica lattice. In the trioctahedral micas the dipole moment of the hydroxyl group is vertical and this places the interlayer potassium ions in a more positive environment than those in the dioctahedral micas where the O–H bond is inclined away from the potassium ion. However, the reexamination of the muscovite structure by Radoslovich (1960) indicates that the bond distance between the potassium ions and the oxygen of the hydroxyl group is too great for the hydroxyl group to have any influence on the stability of the potassium ion.

From the study of the weathering of biotite in a granite from northeast Scotland, Walker (1949) concluded that concomitantly with the release of potassium ions, ferrous ions of the octahedral sheet are oxidized to the ferric state and the net negative charge of the layer is thereby reduced. A mica-vermiculite interlayered mineral forms initially, but with more advanced weathering all the mica is made over to vermiculite. However, where the biotite originates in basic or ultrabasic rocks and the drainage is impeded, montmorillonite tends to be the residual product. MacEwan (1954) has also described a trioctahedral montmorillonite as the breakdown product of biotite. Stephen (1952) noted that in the weathering of biotite-rich primary rocks in Worcestershire, chlorite appears as the early product of weathering of the micas, apparently through the entrapment of released magnesium ions in the interlayer position. The chlorite forms around the margins and along the cleavage planes and spreads inward. Simultaneously, segregations of titania, lime, and silica crystallize as sphene. Higher in the weathered zone, vermiculite replaces chlorite.

According to Wilson (1966), the weathering of biotite in some freely drained soils derived from a biotite-rich, quartz gabbro in Aberdeenshire, Scotland proceeds through expansion of the mineral to yield vermiculite with the simultaneous introduction of a little aluminum to the interlayer positions. With further weathering, the interlayer aluminum content increases, giving rise to mixed layers of vermiculite and chlorite, but subsequently kaolinite and gibbsite develop along the cleavage traces and ultimately may pseudomorph the mineral.

Artificial "weathering" of biotite to vermiculite has been accomplished in the laboratory by Barshad (1954), Mortland (1958), Bassett (1960), and Raussell-Colum et al. (1964), using various inorganic salts.

Muscovite may also weather to a vermiculitelike mineral. Brown (1953) has recorded the presence of an aluminous vermiculite in a soil clay from northwest England and Rich and Obenshain (1955) found a similar mineral in a red-yellow podsolic soil developed on a muscovite schist in the Piedmont Plateau of southeastern United States. In both cases the authors considered the "vermiculite" to have developed from aluminous mica through the loss of potassium ions and their replacement by aluminum.

The accessory minerals of the acid crystalline rocks include the apatites, zircon, and magnetite. Of these, the apatites are of considerable importance since they represent the principal primary source of phosphorus, an element essential to most forms of life. Fluorapatite is the most stable of the group and commonly persists as a detrital mineral in sediments whereas hydroxyapatite is relatively unstable under leaching conditions. The stability of chlorapatite is intermediate between these two. Zirconium is a very insoluble ion and the mineral zircon ($ZrSiO_4$), despite the fact that it is a member of the nesosilicate group, is remarkably stable to chemical weathering. However, Carroll (1953) considered that strongly alkaline conditions might cause partial destruction of the mineral at points of weakness caused by inclusions. Magnetite contains both ferrous and ferric iron and, as such, it is vulnerable to both oxidation and reduction. Oxidation leads to the formation of hematite or maghemite whereas reduction promotes the formation of potentially mobile ferrous compounds such as the bicarbonate. Nevertheless, despite its apparent vulnerability, magnetite is often surprisingly abundant in "heavy-mineral" concentrates of the coarser grained sedimentary rocks.

The Weathering of the Morton Gneiss, Minnesota

In the classical study of the chemical weathering of the Morton granite gneiss, Goldich (1938) obtained six samples of the residual clay from different localities and a "fairly representative" composite sample of the gneiss. These were analyzed by chemical and petrographical means and, from the results, an estimate of the mineralogy was made (Tables 28 and 29). Weathering of

TABLE 28

Chemical Analyses (in Percent) of Fresh and Weathered Morton Gneiss[a]

	Sample[b]						
	1	2	3	4	5	6	7
SiO_2	71.54	69.89	68.09	61.75	70.30	57.53	55.07
Al_2O_3	14.62	16.54	17.31	18.58	18.34	23.57	26.14
Fe_2O_3	0.69	2.33	3.86	1.69	1.55	3.05	3.72
FeO	1.64	0.34	0.36	4.11	0.22	3.59	2.53
MgO	0.77	0.30	0.46	0.76	0.21	0.41	0.33
CaO	2.08	0.06	0.06	0.16	0.10	0.05	0.16
Na_2O	3.84	0.43	0.12	0.10	0.09	0.06	0.05
K_2O	3.92	5.34	3.48	3.54	2.47	0.35	0.14
H_2O+	0.30	4.00	5.14	5.56	5.58	8.82	10.75
H_2O-	0.02	0.35	0.47	0.35	0.30	0.70	0.64
CO_2	0.14	0.21	0.05	1.84	0.20	0.77	0.36
TiO_2	0.26	0.14	0.34	0.92	0.21	0.87	1.03
P_2O_5	0.10	0.04	0.03	0.09	0.04	0.08	0.11
MnO	0.04	0.04	0.06	0.21	0.03	0.07	0.03
BaO	0.09	0.13	0.07	0.09	0.05	Trace	0.01
SO_3	—	Trace	0.00	0.03	0.00	Trace	Trace
S	0.02	0.01	0.01	0.05	0.01	0.01	0.04
Total	100.07	100.15	99.91	99.83	99.70	99.93	100.11

[a]After Goldich (1938); used by permission of The University of Chicago Press.
[b]Sample 1: representative composite of the Morton granite gneiss. Samples 2–7: weathered clay derived from the granite gneiss.

the gneiss had taken place prior to the Cretaceous and the residual clay at present is covered in part by glacial drift.

The essential constituents of the gneiss are zoned oligoclase, orthoclase, microcline, quartz, biotite, and hornblende, while magnetite, apatite, zircon, sphene, and allanite are present as accessories only. According to Goldich, the order of destruction of these minerals is as follows:

Least stable
 Plagioclase, epidote,
 hornblende, sphene,
 apatite
Moderately stable
 Ilmenite/magnetite,
 Biotite
Most stable
 Microcline, orthoclase,
 zircon, quartz

Since the samples were not obtained from a single, vertical weathered sequence, the degree of weathering to which each had been subjected cannot be determined with certainty. However, if it be assumed that the gneiss were

TABLE 29

Mineralogical Composition (in Percent) of the Morton Gneiss and Residual Clays[a]

	Sample [b]						
	1	2	3	4	5	6	7
Quartz	30.0	35.0	40.0	31.0	43.0	28.0	25.0
K-Feldspar	19.0	31.0	18.0	19.0	13.0	2.0(?)	1.0(?)
Plagioclase	40.0	4.0	1.0	1.0	1.0	?	?
Biotite	7.0	0.8	Trace	4.0	Trace	0.5	0.2
Hornblende	1.0	Trace	0.0	Trace	0.0	0.1	Trace
Magnetite	1.0	—	—	—	—	—	—
Ilmenite	0.5	3.0	5.0	4.0	2.0	6.0	6.0
Leucoxene, etc.	Trace	—	—	—	—	—	—
Calcite	0.3	—	—	—	—	—	—
Siderite	—	0.6	0.1	5.0	0.5	2.0	0.9
Sphene	0.04	0.0	0.0	Trace	Trace	0.0	0.0
Epidote	0.2	Trace	Trace	0.2	Trace	Trace	Trace
Apatite	0.2	Trace	Trace	Trace	Trace	Trace	Trace
Zircon	0.02	0.02	0.03	0.08	0.02	0.15	0.10
Allanite	Trace	—	—	—	—	—	—
Kaolin	—	26.0	36.0	35.0	40.0	59.0	66.0

[a] After Goldich (1938); used by permission of The University of Chicago Press.
[b] Sample 1: representative composite of the Morton granite gneiss. Samples 2–7: weathered clay derived from the granite gneiss.

homogeneous throughout the sampling area and that alumina remained immobile during the weathering process, a comparison of the alumina content of each sample of residual clay, recalculated on an anhydrous basis, with that of the unaltered parent rock should prove indicative of the order of weathering of the samples. In Fig. 47, the concentration ratios for the various constituents have been calculated from Goldich's data and adjusted so that alumina has unit concentration ratio throughout. The values so obtained have been plotted against the ratio of the alumina content of each sample to that of the parent gneiss, as the ordinate. The TiO_2 and MgO values for sample 5 proved erratic and, for convenience, have been disregarded. Figure 48 is a graphical representative of the mineralogical data relative to the same ordinate. A comparison of these two diagrams shows that during the early stages of alteration there is a marked loss of soda and lime and a slight increase in potash concentration, a trend undoubtedly brought about by the differential destruction of the plagioclase relative to the potash feldspars. However, with increasing weathering the potash feldspars are gradually converted to kaolin also. The silica curve shows a steady decrease in concentration of this constituent as the weathering intensity becomes more pronounced. This is brought about mainly through the conversion of the feldspars to kaolin, but in the more weathered areas some loss of quartz also occurs. The iron oxides and titania concentration ratios have sympathetic

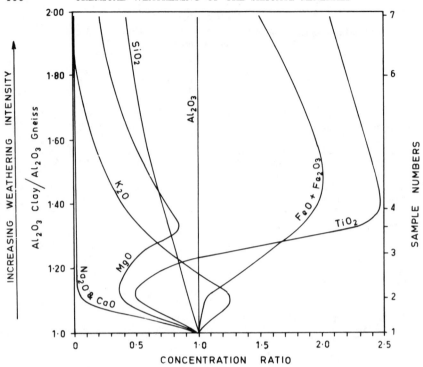

Figure 47. Concentration ratios for various oxides in weathered clays derived from the Morton gneiss in relation to the degree of weathering (data from Goldich, 1938).

trends, increasing to a maximum in sample 4 and then decreasing gradually toward the top of the diagram.

Bauxite Laterites Developed on Granite in Western Australia

Grubb (1966) has described the texture and mineralogy of a number of bauxitic laterite sequences developed from granite in Western Australia. The alteration is believed to have taken place following a period of peneplanation during the Miocene and subsequently was affected by an epeirogenic uplift. The lateritic surfaces, at present, vary in elevation from 200 to 1850 feet above sea level. Although the investigations were not continued to the parent rock, as shown in Fig. 49, the mineralogical variations with depth are not appreciably different from those at Weipa (Fig. 53), where the parent material is a kaolinitic sand. Kaolinite or halloysite or both were the earliest minerals to form, but as the leaching intensified upward, partial desilification caused a reduction in the quartz content and converted much of the kaolinitic clay to bauxite minerals. Iron apparently migrated upward from the bleached clay zone to the surface horizons where it was stabilized through oxidation, yielding goethite.

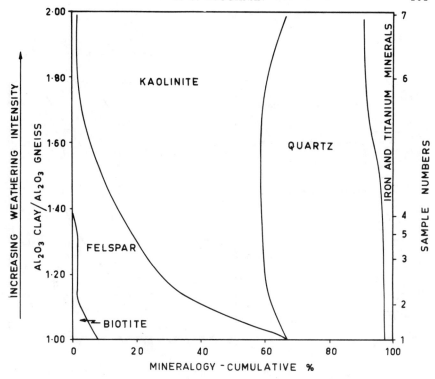

Figure 48.   Mineralogy in relation to degree of weathering of the Morton gneiss (data from Goldich, 1938).

ALKALINE IGNEOUS ROCKS

The alkaline igneous rocks are characterized by a high alkali content in relation to silica and (or) alumina when compared with normal igneous rocks and this high alkali-to-silica ratio finds expression in the development of minerals such as the feldspathoids. Although these rocks are not particularly abundant, the instability of some of their constituents and the fact that a few of the world's bauxite deposits have resulted from their decomposition renders them of particular interest.

The feldspathoids, nepheline ($NaAlSiO_4$), and the less common leucite ($KAlSi_2O_6$), have framework structures resembling the alkali feldspars but, unlike the latter, they are particularly unstable in the weathering environment and are among the earliest minerals to be destroyed. Dorfman (1958) has shown that nepheline hydrolyzes rapidly in distilled water yielding equilibrium pH values of the order of 11. Moreover, Keller *et al.* (1963), in a study of the decomposition of minerals by fine grinding, found that the solutions in contact with pulverized nepheline contained the highest contents of dissolved salts of any of the minerals examined.

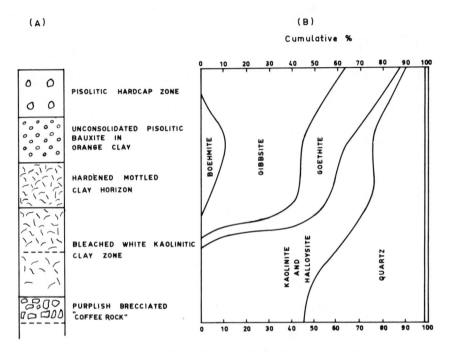

Figure 49.   Mineralogical variations with depth in the Angus Cut profile developed on granite, Western Australia (after Grubb, 1966).(A) The weathered sequences. (B) Cumulative percentage by weight of the main mineral components.

Weathering of Nepheline Syenites in Brazil

Harder (1952) has described the weathering of phonolite and nepheline syenite to massive bauxite deposits in the Poços de Caldas district of Brazil (Table 30 and Fig. 50). The nepheline-rich rocks occur as intrusive masses of limited extent in Pre-Cambrian granite and gneiss. The residual layer of porous, cellular bauxite extends to depths of 50 feet or more and is separated from the unaltered rock by a sharp although extremely irregular contact. The sharpness of the contact is emphasized where a single crystal may be completely fresh at one end below the contact, yet altered to gibbsite at the other above the contact. Unaltered boulders of nepheline syenite, varying in size down to a matter of a few inches, occur within the bauxite. Also contained within the porous residual bauxite are frequent lenticular masses of white, red, or varicolored kaolinitic clay and in places the same material occurs between the bauxite and the unaltered nepheline syenite. The clay apparently represents the decomposed remnants of more siliceous dykes, sills, or xenoliths within the original rock mass. The weathering environment, although sufficiently intense to cause the complete alteration of the nepheline to bauxite has been unable to bring about desilification of these kaolinitic masses.

TABLE 30

Chemical Analyses (in Percent) of the Alkaline Rocks and Derived Bauxites from Poços de Caldas, Brazil[a]

	Phonolite	Foyaite	Bauxite	
			Porous residual	Surface fragmental
SiO_2	53.06	53.18	5.50	1.60
Al_2O_3	20.85	21.18	55.06	58.58
Fe_2O_3	4.47	3.39	9.60	6.87
FeO	0.79	0.76	—	—
MgO	0.61	1.60	0.008	0.008
CaO	1.29	1.01	0.02	0.02
Na_2O	7.18	7.34	—	—
K_2O	8.32	8.42	—	—
H_2O+	1.67	1.44	27.65	31.50
H_2O-	0.25	1.20	—	—
TiO_2	0.35	0.70	1.77	1.11
P_2O_5	0.10	1.10	0.33	0.18
BaO	0.20	0.18	—	—
MnO	—	—	0.06	0.13

[a] After Harder (1952).

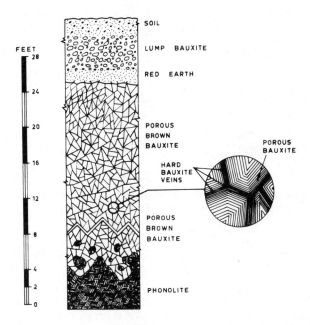

Figure 50.   Vertical section through the bauxite deposit at Poços de Caldas, Brazil (after Harder, 1952).

Weathering of Nepheline Syenite in Arkansas

In the Little Rock area of Arkansas, the weathering of nepheline syenite during the early Tertiary has resulted in one of the world's major bauxite deposits.

According to Gordon and Tracey (1952) the nepheline syenite, which intruded folded Palaeozoic sediments probably during the late Cretaceous, was exposed by erosion in the Palaeocene. Weathering produced a deep profile in which the nepheline syenite grades upward into a completely kaolinized zone still retaining in part the original texture of the parent rock. A zone of compact kaolinitic clay separates the kaolinized nepheline syenite from the overlying massive, granite-textured bauxite. Overlying the granite-textured bauxite and separated from it by a sharp but irregular contact is a zone of pisolitic and vermicular bauxite (Fig. 51).

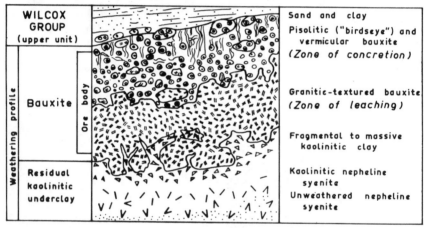

Figure 51.   Diagrammatic section illustrating weathering of nepheline syenite in Arkansas (after Gordon and Tracey, 1952).

An interesting feature of these deposits is that the bauxite formed from the nepheline syenite on moderate to steep slopes, which in most places have dips of more than 5 degrees and generally range from 10 to 25 degrees. Apparently the rate of decomposition of the nepheline syenite and the formation of bauxite were sufficiently rapid to overcome erosion from these steep slopes.

ARGILLACEOUS SEDIMENTS

For convenience, it is proposed to group together, irrespective of their textures and structures, all sedimentary rocks which contain clay minerals and whose weathering is characterized by the alteration of these minerals. Included, therefore, are not only the shales and claystones but also argillaceous sands and sandstones, loess, and till deposits as well as low-grade metamorphics such as argillites and slates. Apart from the clay minerals, these rocks almost invariably contain quartz, generally as a major constituent, while feldspars and "heavy minerals" such as hematite, rutile, zircon, tourmaline, and so on, and carbonates, commonly occur but in small amounts only.

The mechanisms of weathering of the clay minerals have been the subject of a considerable number of investigations over the past two decades and a great deal of information has emerged. Unfortunately, however, much of the data are difficult to present in simple terms because, upon breakdown, the clay minerals tend to form metastable intermediate phases which frequently are definable only in terms of their X-ray data and then only after prior treatment with cations, such as K^+ and Mg^{++}, and various polar liquids.

The illites are, by far, the most abundant of the clay minerals in argillaceous rocks. They occur in fine- and coarse-grained sediments of both marine and freshwater origin and in the low-grade metamorphic derivatives of these rocks. Dioctahedral aluminous types predominate although locally glauconite may be abundant and trioctahedral magnesium and ferrous illites are known. The aluminous illites weather in much the same manner as muscovite mica although their finer grain size and lower degree of crystallinity render them more vulnerable to the leaching solutions. Loughnan et al. (1962) have described the weathering of aluminous illite in shales of Triassic age from the Sydney area of New South Wales. These rocks have been subjected to lateritization which is generally believed to have taken place during the Tertiary although it is possible the processes are continuing at present. Proceeding from the parent shales upward to the lateritic surface (Fig. 52), little alteration

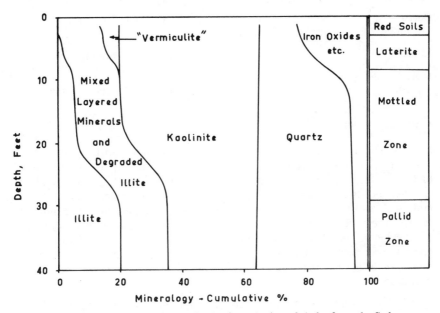

Figure 52.   Mineralogy in relation to depth of weathering of shales from the Sydney area, Australia (data from Loughnan et al., 1962).

of the illite takes place until the mottled zone is reached. Within this zone, however, potassium is leached from the mineral and the highly charged residual structure is free to expand in a manner similar to montmorillonite. In this form the mineral is particularly vulnerable to desilification and conversion to kaolinite. Nevertheless, some of the degraded mineral persists through to the lateritic zone where sorption of ferric or aluminum ions or both stabilizes the structure with the formation of a 14 Å mineral, which resembles vermiculite in that it shows little expansion on treatment with polar liquids yet collapses to 10 Å after heating to 450°C. Rolfe and Jefferies (1953) and Rich and Obenshain (1955) have described the occurrence of similar "vermiculitelike" minerals in soils from eastern United States.

In the study of the chemical weathering of glauconite at Gingin, Western Australia, Cole (1940) found the initial alteration product to be a member of the montmorillonite group, but in the soil zone this mineral becomes unstable and is converted to kaolinite (Table 31). Wurman (1960), on the other hand,

TABLE 31

Weathering of Glauconite at Gingin, Western Australia[a]

Mineral	"Upper greensand"		Subsoil		Soil	
	Coarse	Fine	Coarse	Fine	Coarse	Fine
Glauconite	Very much	Much	Little	Little	Possible trace	Possible trace
Quartz	Little	Very little	Much	Very little	Very much	—
Montmorillonite	Possible trace	Possible trace	Much	Much	—	—
Kaolinite	—	—	Little	Little	Possible trace	Much
Hematite and/or goethite	Very little	—	—	Possible trace	Little	Little

[a] After Cole (1940).

has described the weathering of glauconite in Cambrian sandstones from Wisconsin to yield highly ferruginous soils. The alteration apparently proceeded in two stages, (a) the loss of potassium and iron from the lattice and the expansion of the glauconite to yield a randomly interstratified system of 10 Å and 14 Å and (or) 17 Å and (b) the breakdown of the interlayer minerals leaving a residue of oxides and hydroxides of iron. According to Wolff (1967) weathering of the Aquia Greensand in Maryland results in the formation of goethite pseudomorphs after the glauconite and the secondary precipitation of kaolinite. He has proposed the following reaction scheme to account for the alteration.

$$H_2O + \text{glauconite} \rightarrow \text{goethite} + \text{dissolved matter} + H_2O$$
$$\xrightarrow{\text{precipitation}} \text{kaolinite} + \text{dissolved matter} + H_2O.$$

Chlorites also occur in both marine and freshwater sedimentary rocks, although generally they are present as minor constituents and commonly they are interlayered with other clay minerals. A notable exception to this trend has been described recently by Ball (1966) from North Wales, where both the parent volcanic ash and the derived soils are rich in chlorite minerals. Droste and Tharin (1958) have given an interesting account of the mechanism of weathering of chlorite in till of Illinoian age from Pennsylvania. In the early stages of alteration "hydroxyls of the brucite sheet are joined by protons slowly, steadily and randomly but no 'islands' of Mg^{++} regularly co-ordinated with H_2O develop within the brucite sheet." Ultimately, however, all the hydroxyls of the brucite sheet are converted to water and the reorganization of the magnesium and water results in the formation of vermiculite. The talc layers of the original chlorite are apparently unaffected by the modifications to the brucite sheet and are inherited by the vermiculite.

Bayliss and Loughnan (1964) have drawn attention to the considerable disparity in weathering stability between chlorite and talc. In a thick weathered sequence developed on a parent material containing both chlorite and talc near Orange, New South Wales, they found that whereas chlorite is rapidly destroyed near the base of the sequence, talc is considerably more resistant and persists through to the surface horizon. They considered the brucite sheet as the weakness in the chlorite structure. With its destruction, the residual "talc" layer of the chlorite is left with a high charge. Hydrogen ions are attracted to the surface and in this form the structure is particularly vulnerable to further breakdown. Talc, on the other hand, has a balanced structure and neither hydrates readily nor attracts hydrogen ions to its surface.

Although montmorillonites are relatively common as initial decomposition products of volcanic glasses and primary minerals such as pyroxenes, amphiboles, and plagioclases, they are not particularly abundant in sedimentary rocks. Probably the reason for the discrepancy is that the montmorillonites carry a net charge and hence are susceptible to alteration under leaching conditions. However, there appears to be some disagreement concerning the actual weathering stability of montmorillonites. Jackson et al. (1948), for example, placed the montmorillonites after the micas and illites in the weathering sequence of minerals (Table 12) whereas Bayliss and Loughnan (1964), in a study of the weathering of clay slates from New South Wales, found illite considerably more resistant than montmorillonite to chemical breakdown. The confusion has probably arisen through failure to differentiate between montmorillonites and illites which have been depleted of their potassium ions.

The weathering of ferric-aluminous montmorillonite to form an assemblage of kaolinite and goethite has been described by Altschuler and Dwornik (1963). According to these authors the transformation is accomplished through leaching of interlayer cations and tetrahedral silica layers; initially

this creates a regular 1 : 1 mixed layer of montmorillonite-kaolinite, but ultimately this structure is replaced by hexagonal plates of kaolinite. The iron
released during these reactions crystallizes as goethite while the silica enriches
the ground waters and gives rise to secondary deposits of chert.

Kaolinite is of widespread distribution in sedimentary rocks. According
to Weaver (1958), it is most abundant in near-shore and continental, particularly fluviatile, environments and is less common in ancient marine sediments. It is generally absent from salt lake deposits. In the weathering
environment, kaolinite is the most stable of the clay minerals and forms the
penultimate phase in the weathering sequences of alumino-silicate minerals.
However, under conditions of intense leaching, silica is lost from the lattice
and the residual alumina crystallizes as gibbsite.

$$\underset{\text{kaolinite}}{Al_2Si_2O_5(OH)_4} + \underset{\text{water}}{H_2O} \rightleftarrows \underset{\text{gibbsite}}{2Al(OH)_3} + \underset{\text{silica}}{2SiO_2}$$

This reaction appears to be completely reversible and the virtual absence
of gibbsite in rocks older than the Tertiary is probably attributable to the
marked affinity of the gibbsite structure for silica. Partial dehydration of
gibbsite leads to the development of the monohydrate boehmite (and possibly diaspore), in which form the alumina is very unreactive.

Dickite, nacrite, and halloysite are either rare or unknown in sedimentary
rocks.

Weathering of Loess and Till in the Upper Mississippi Valley

In a study of weathered profiles developed on till and loess in Indiana,
Droste *et al.* (1960) found illite and lesser amounts of chlorite present in the
parent materials. Upon weathering, both minerals tend to lose their interlayer ions and form expandable, degraded structures which resemble montmorillonite. Chlorite is the first to be attacked and in all profiles studied it
shows the greater degree of alteration. The authors considered the higher
kaolinite contents in the upper levels of the weathered zones could have
resulted from either unequal distribution of the mineral in the parent material or alteration of clay and other silicate minerals.

Glenn *et al.* (1959) described the weathering of loess in a prairie soil
profile in southwestern Wisconsin. Ferruginous montmorillonite proved to
be the dominant clay mineral in the parent loess although illite and vermiculite are relatively abundant and kaolinite, amorphous material, and chlorite
form minor constituents. The weathering sequence has been obscured by
the downward migration of some of the montmorillonite (Fig. 53). Nevertheless, there is evidence indicating that partial destruction of the montmorillonite and vermiculite has occurred and that this trend has been accompanied by a progressive increase in the kaolinite, chlorite, and amorphous
silica and alumina contents. The weathering stability displayed by chlorite
in this profile is in contrast to its behavior observed elsewhere. The authors
suggest that interlayered chlorite-vermiculite forms an intermediate phase
in the transition of montmorillonite to chlorite.

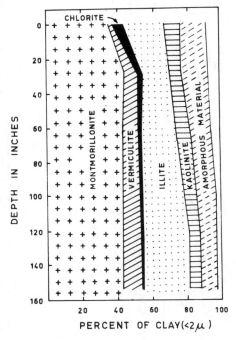

Figure 53.   Mineralogy expressed as a percentage of the $<2\ \mu$ clay fraction in relation to
depth in the Tama silt loam, Wisconsin (after Glenn *et al.*, 1959).

Weathering of Clay-Slates in New South Wales

Bayliss and Loughnan (1963) have described a number of weathered
sequences developed on Palaeozoic clay-slates in New South Wales. The
rainfall in the areas studied ranges from 21 to 46 inches per annum and the
average mean temperatures from 57° to 64°F, while the topography varies
from essentially flat to gently undulating.

Quartz and well-crystallized illite (2M₁ polymorph) form the dominant
constituents of the clay-slates although montmorillonite and chlorite, the
latter interlayered with expandable lattice material, are invariably present.
Weathering has produced a deep profile (Fig. 54) in which the steeply folded
slates grade upward into a thick, structureless leached zone and this in turn
is overlain by a reddish ferruginous zone of accumulation. The upper and
lower boundaries of the leached zone approximate the limits of the fluctuating water table. Chlorite is the least stable of the primary constituents and
is destroyed near the base of the leached zone. Montmorillonite is more
resistant to the percolating waters and persists into the ferruginous zone
whereas degradation and destruction of the illite is restricted to the surface
layer only. Kaolinite is the end product of the breakdown of these clay
minerals.

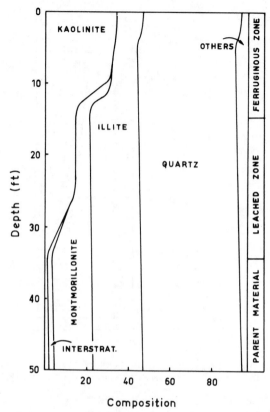

Figure 54.   Mineralogy in relation to depth for the weathering of clay slates at Goulburn,
New South Wales (after Bayliss and Loughnan, 1964).

Weathering of Kaolinitic Sandstone at Weipa, Queensland

One of the most spectacular examples of the chemical weathering of sedimentary rocks is that in evidence at Weipa in northern Queensland (Loughnan and Bayliss, 1961). Here, under a hot monsoonal climate with a mean
temperature of 82°F and an average rainfall of 60 inches per annum, sandstone consisting of nearly 90% quartz and 10% kaolinite has been leached
of the bulk of its silica and has given rise to thick deposits of lateritic bauxite
(Fig. 55 and Table 32).

The parent sandstone is loosely consolidated and contains coarse angular
quartz with a kaolinitic matrix. This grades upward into a thick, reddish,
friable to compact zone, representing the limits of the fluctuating water
table. In the lower few feet of this zone, quartz and kaolinite are the principal constituents but in the relatively more arid upper regions, gibbsite
predominates. The coarse, angular characteristics of the quartz prevail
throughout while etched and pitted surfaces are apparent in some of the

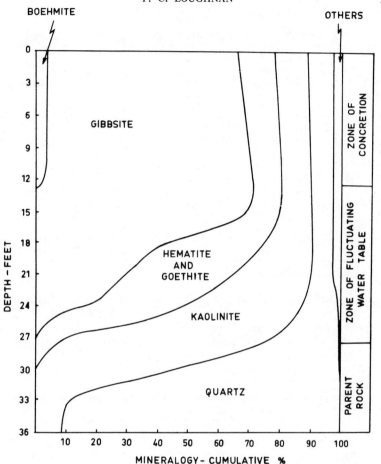

Figure 55.   Mineralogy in relation to depth at Weipa, Queensland (after Loughnan and
Bayliss, 1961).

grains. Hematite and goethite, mainly in the form of irregular concretions, attain their maximum development in this zone.

The uppermost concretionary zone is clearly defined from the underlying material and consists of gibbsite, hematite, and (or) goethite in the form of roughly spherical pisolites with minor amounts of kaolinite, quartz, and boehmite.

Fossil Bauxite Developed on Clay Slate near Portilla de Luna, Spain

Font-Altaba and Closas M. (1960) have recorded the occurrence of an interesting fossil bauxite in Devonian strata near Portilla de Luna, Spain. The material consists of a white soft clay, composed of gibbsite (54–58%) and fully hydrated halloysite (32–38%) with blebs of coal and minor amounts of carbonates and iron oxide. The bauxite is overlain by a fossiliferous lime-

TABLE 32

Chemical Composition of the Weathering Sequence at Weipa, Queensland[a]

Sample	Depth (feet)	Analysis (%)								
		SiO_2	Al_2O_3	Fe_2O_3	TiO_2	K_2O	Na_2O	H_2O+	H_2O-	Total
W1	1–4	9.9	47.1	13.1	2.1	—	—	24.6	2.2	99.0
W2	4–7	10.0	46.9	12.9	2.1	—	—	24.5	2.2	98.6
W3	7–10	10.6	46.8	11.5	2.2	—	—	24.5	2.2	99.8
W4	10–13	10.5	49.4	9.6	2.2	—	—	25.3	2.3	99.3
W5	13–16	9.6	50.0	9.2	2.2	—	—	26.6	2.1	99.7
W6	16–19	16.9	36.1	25.4	1.2	—	—	19.1	1.4	100.1
W7	19–22	19.4	34.3	26.3	1.2	—	—	16.6	1.2	99.0
W8	22–25	26.5	31.5	26.1	0.9	—	—	13.4	1.2	99.6
W9	25–28	31.5	28.6	25.8	0.9	—	—	11.4	1.0	99.2
W10	28–29	73.1	17.9	2.8	0.7	—	—	5.6	0.6	100.7
W11	31–32	92.9	4.1	0.5	0.2	—	—	1.0	0.0	98.7
W12	33–34	91.1	5.2	0.5	0.3	0.2	0.7	1.8	0.0	99.8
W13	35–36	93.2	3.9	0.2	0.2	0.1	0.8	1.4	0.0	99.8

[a] After Loughnan and Bayliss (1961).

stone and underlain by fragile clay slates which, the authors believe, represent the parent material. Partial chemical analyses of the clay slates and three samples of the bauxite are shown in Table 33. No explanation is given for the paucity of titania in the bauxite. This is one of the few recorded occurrences of gibbsite in rocks older than the Tertiary.

TABLE 33

Chemical Analyses for the Bauxites near Portilla de Luna, Spain and Underlying Clay Slates[a]
Bauxite[b] analysis (%)

	1	2	3	4
SiO_2	44.10	13.20	14.26	15.43
Al_2O_3	27.43	49.70	44.61	47.92
Fe_2O_3	11.63	1.10	0.35	0.28
TiO_2	0.90	Trace	Trace	Trace
CaO	—	1.39	1.37	1.29
MgO	—	2.44	3.08	0.42
Wt. loss (at 800°C)	11.97	32.95	36.87	36.55
Total	96.03	100.78	100.54	101.89

[a] After Font-Altaba and Closas M. (1960).
[b] (1) Parent clay slate; (2) yellowish bauxite, (3) white bauxite; (4) white bauxite.

REFERENCES

Alexander, L. T., S. B. Hendricks, and G. T. Faust (1941), Occurrence of gibbsite in some soil forming materials, *Proc. Soil Soc. Am.* **6**, 52–57.

Altschuler, Z. S., and E. J. Dwornik (1963), Transformation of montmorillonite to kaolinite during weathering, *Science* **141**, 148–152.

Ball, D. F. (1966), Chlorite clay minerals in Ordovician pumice-tuff and derived soils in Snowdonia, North Wales, *Clay Min. Bull.* **6**, 195–210.

Barshad, I. (1954), Cation exchange in micaceous minerals. II. Replaceability of ammonium and potassium from vermiculite, biotite and montmorillonite, *Soil Sci.* **78**, 57–76.

Bassett, W. A. (1960), Role of hydroxyl orientation in mica alteration, *Bull. Geol. Soc. Am.* **71**, 449–456.

Bates, T. F. (1952), Interrelationships of structure and genesis in the kaolinite group. Problems of clay and laterite genesis. *Am. Inst. Min. Met.*, pp. 144–153.

Bates, T. F. (1962), Halloysite and gibbsite formation in Hawaii, *Proc. Nat. Conf. Clays and Clay Minerals* **9**, 307–314.

Bayliss, P., and F. C. Loughnan (1963), Mineralogical evidence for the penecontemporaneous lateritization of basalts from New England, N.S.W., *Am. Mineralogist* **48**, 410–414.

Bayliss, P. S., and F. C. Loughnan (1964), Mineralogical transformations accompanying the chemical weathering of clay slates from New South Wales, *Clay Min. Bull.* **5**, 353–362.

Brown, G. (1953), The dioctahedral analogue of vermiculite, *Clay Min. Bull.* **2**, 64–69.

Carrol, D. (1953), Weatherability of zircon, *J. Sed. Petrol.* **23**, 106–116.

Chapman, R. W., and M. A. Greenfield (1949), Spheroidal weathering of igneous rocks, *Am. J. Sci.* **247**, 407–429.

Cole, W. F. (1940), X-ray analysis and microscopic examination of the product of weathering of the Gingin upper greensand, *J. Roy. Soc. W. Australia* **27**, 229–243.

Craig, D. C. (1963), Geochemical and mineralogical aspects of the weathering of some basaltic rocks from New South Wales. Unpubl. M. Sc. Thesis, Univ. New South Wales.

Craig, D. C., and F. C. Loughnan (1964), Chemical and mineralogical transformations accompanying the weathering of basic volcanic rocks from New South Wales, *Australian J. Soil Res.* **2**, 218–234.

De Vore, G. W. (1958), The surface chemistry of feldspars as an influence on their decomposition products, *Proc. Nat. Conf. Clays and Clay Minerals* **6**, 26–41

Dorfman, M. D. (1958), Geochemical characteristics of weathering processes in nepheline syenites of Khibina tundra, *Geochem. Trans. Reoxmma* **5**, 537–551.

Droste, J. B., and J. C. Tharin (1958), Alteration of clay minerals in Illinoian till by weathering, *Bull. Geol. Soc. Am.* **69**, 61–68.

Droste, J. B., N. Bhattacharya, and J. A. Sunderman (1960), Clay mineral alteration in some Indiana soils, *Proc. Nat. Conf. Clays and Clay Minerals* **9**, 329–342.

Eyles, V. A. (1952), The composition and origin of the Antrim laterites and bauxites, *Mem. Geol. Surv. N. Ireland*.

Font-Altaba, M., and M. J. Closas (1960), A bauxite deposit in the paleozoic of Leon, Spain, *Econ. Geol.* **55**, 1285–1290.

Glenn, R. C., M. L. Jackson, F. D. Hole, and G. B. Lee (1960), Chemical weathering of layer silicate clays in loess-derived Tama silt loam of southwestern Wisconsin, *Proc. Nat. Conf. Clays and Clay Minerals* **8**, 63–83.

Goldich, S. S. (1938), A study of rock weathering, *J. Geol.* **46**, 17–58.

Gordon, M., and J. I. Tracey (1952), Origin of the Arkansas bauxite deposits. Problems of clay and laterite genesis, *Am. Inst. Min. Met.*, pp. 12–34.

Grant, W. H. (1963), Chemical weathering of biotite-plagioclase gneiss, in *Proc. Nat. Conf. Clays and Clay Minerals* **12**, 455–463.

Grubb, P. L. C. (1966), Some aspsects of lateritization in Western Australia, *J. Roy. Soc. W. Australia* **49**, 117–124.

Harder, E. C. (1952), Examples of bauxite deposits illustrating variations in origin. Problems of clay and laterite genesis, *Am. Inst. Min. Met.*, pp. 55–64.

Hay, R. L. (1959), Origin and weathering of late Pleistocene ash deposits on St. Vincent, B.W.I., *J. Geol.* **67**, 65–87.

Jackson, M. L., S. A. Tyler, A. L. Willis, G. A. Bourbeau, and P. Pennington (1948), Weathering sequence of clay-size minerals in soils and sediments. II. Chemical weathering of layer silicates, *Proc. Soil Sci. Am.* **16,** 3–6.

Jenny, H., and C. D. Leonard (1934), Functional relationships between soil properties and rainfall, *Soil Sci.* **38,** 363–381.

Keller, W. D., W. D. Balgord, and A. L. Reesman (1963), Dissolved products of artificially pulverized silicate minerals and rocks, Pt. 1, *J. Sed. Petrol.* **33,** 191–204.

Krauskopf, K. B. (1959), The geochemistry of silica in sedimentary environments. Silica in sediments, *Soc. Econ. Paleontologists-Mineralogists Symp.* **7,** 4–18.

Loughnan, F. C., and P. Bayliss (1961), The mineralogy of the bauxite deposits near Weipa, Queensland, *Am. Mineralogist* **46,** 209–217.

Loughnan, F. C., and P. Bayliss (1963), A chromium-bearing dyke clay from Cowan, N.S.W., *Australian J. Sci.* **26,** 185.

Loughnan, F. C., and H. G. Golding (1957), The mineralogy of the commercial dyke clays in the Sydney district, *J. Roy. Soc. New South Wales* **91,** 85–91.

Loughnan, F. C., R. E. Grim, and J. Vernet (1962), Weathering of some Triassic shales in the Sydney area, *J. Geol. Soc. Australia* **8,** 245–258.

MacEwan, D. M. C. (1954), "Cardenite," a trioctahedral montmorillonite derived from biotite, *Clay Min. Bull.* **2,** 120–125.

McKenzie, R. C., G. F. Walker, and R. Hart (1949), Illite occurring in decomposed granite at Ballater, Aberdeenshire, *Min. Mag.* **28,** 704–713.

Radoslovich, E. V. (1960), The structure of muscovite, *Acta Cryst.* **13,** 919–932.

Rausell-Colom, J. A., T. R. Sweatman, C. B. Wells, and K. Norrish (1964), Studies in the artificial weathering of mica, *Exptl. Pedol.*, 11th East. School, Nottingham, pp. 40–72.

Rich, C. I., and S. S. Obenshain (1955), Chemical and clay mineral properties of a red and yellow podsolic soil derived from muscovite schist, *Proc. Am. Soil Soc.* **19,** 334–339.

Sand, L. B. (1956), On the genesis of residual kaolins, *Am. Mineralogist* **41,** 28–40.

Sherman, G. D. (1949), Factors influencing the development of lateritic and laterite soils in the Hawaiian Islands, *Pac. Sci.* **3,** 307–314.

Sherman, G. D. (1952a), The titanium content of Hawaiian soils and its significance, *Proc Am. Soil Soc.* **16,** 15–18.

Sherman, G. D. (1952b), The genesis and morphology of the alumina-rich laterite clays. Clay and laterite genesis, *Am. Inst. Min. Met.*, pp. 154–161.

Sherman, G. D., H. Ikawa, G. Uehara, and E. Okasaki (1962), Types of occurrence of nontronite and nontronite-like minerals in soils, *Pac. Sci.* **16,** 57–62.

Siever, R. (1957), The silica budget in the sedimentary cycle, *Am. Mineralogist* **42,** 821–841.

Smith, W. W. (1962), Weathering of some Scottish basic igneous rocks with reference to soil formation, *J. Soil Sci.* **13,** 202–215.

Stephen, I. (1952), A study of rock weathering with reference to the soils of the Malvern Hills, *J. Soil Sci.* **3,** 20–33.

Stephen, I. (1963), Bauxite weathering at Mount Zomba, Nyasaland, *Clay Min. Bull.* **5,** 203–208.

Von Schellmann, W. (1964), Zur lateritischen Verwitterung von Serpentinit, *Geol. Jahrb.* Hannover **81,** 645–678.

Walker, G. F. (1949), Decomposition of biotite in the soil, *Min. Mag.* **28,** 693–708.

Weaver, C. E. (1958), Geologic interpretation of argillaceous sediments. Pt I. Origin of clay minerals in sedimentary rocks, *Bull. Am. Assoc. Petrol. Geol.* **42,** 254–272.

White, H. P., and J. C. Mingaye (1904), *Rec. Geol. Surv. New South Wales* **7,** 230.

Wilson, M. J. (1966), The weathering of biotite in some Aberdeenshire soils, *Min. Mag.* **35,** 1080–1093.

Wolff, R. G. (1967), X-ray and chemical study of weathering glauconite, *Am. Mineralogist* **52,** 1129–1138.

Wurman, E. (1960), A mineralogical study of a gray brown podsolic soil in Wisconsin derived from glauconitic sandstone, *Soil Sci.* **89,** 38–44.

VI. Chemical Weathering and Soil Formation

INTRODUCTION

Soil science or pedology (Greek, *pedon*=ground, *logia*=knowledge) had its beginning toward the latter part of the last century mainly through the advances made by the Russian school led by Dokuchaiev and, later, Glinka. While fully appreciating the influence exerted by the parent rock, this school emphasized the role of the environmental factors, particularly climate, topography, and biologic agencies, in the development of soil characteristics. In modern pedology soils are recognized as representing more than just "the upper weathered layer of the earth's crust" (Ramann, 1911); rather, they are comprised of a complex system of air, water, decomposing organic matter, living plants, and animals in addition to the residues of rock weathering, organized into definite structural patterns as dictated by the environmental conditions. However, in the development of a specific soil, generally not one but several environments prevail, superimposed upon one another, giving rise to differentiated zones or *horizons*. These horizons contrast with those above and (or) below in composition, texture, structure, and (or) color, and are often separated from one another by sharply defined boundaries. Extending from the surface down toward the unaltered parent rock, the horizons were designated by Dokuchaiev as A, B, C, and D in that order and the complete succession is termed the *soil profile*. A generalized profile is shown in Fig. 56.

The A or surface horizon is defined as the zone of *eluviation*, that is, the zone which has been depleted of material carried downward either in suspension or solution, by percolating waters. The underlying B horizon, in contrast, is the zone of *illuviation*, that is, the zone of deposition of materials derived from the A horizon. In actual practice these definitions may be difficult to apply. Thus, for example, the A horizon may receive considerable amounts of mineral matter introduced by surface wash or wind or from the residue of decaying plant debris, while despite enrichment from above, the B horizon is nevertheless composed of highly weathered material which has been leached of most of the original soluble constituents so that it bears little resemblance to the parent rock.

Probably the most striking example of these eluviation-illuviation processes is afforded by the development of lime-rich or *caliche* layers in the soils of arid and semiarid regions. The lime occurs as nodules, generally about pea sized, although they may range up to a foot or more in length, or as disseminated flecks and particles. The nodules may be cemented into essentially continuous layers. Jenny (1950) has offered the following explanation for the origin of these lime layers:

"If a uniform parent material containing some calcium carbonate is

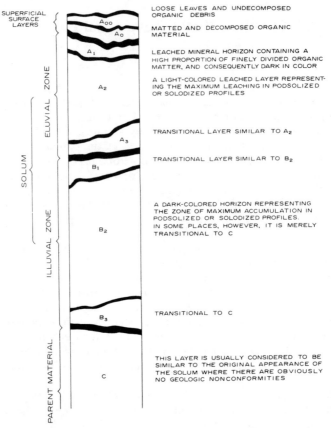

Figure 56. A generalized profile of a timbered soil developed in a humid climate with a
moderate temperature (after Millar, Turk, and Foth, 1958).

assumed, the formation of the lime horizon may be visualized as the consequence of calcium carbonate-bicarbonate equilibria which are regulated by
the carbon dioxide pressure of soil air.

"Root respiration and decay of vegetable matter which are very active in
the surface soil produce large amounts of carbon dioxide (according to
Keller, 1957, this may be ten times that of the normal atmosphere). It converts the relatively insoluble calcium carbonate to the much more soluble
calcium bicarbonate. Percolating rain water translocates the bicarbonates
from the surface to the subsoil. There, owing to reduced CO_2 pressure of the
soil air, which is the result of a low biological activity, calcium carbonate is
precipitated as lime concretions. In areas of low rainfall, the carbonate horizon is close to the surface. As annual rainfall increases, the lime moves to
greater depth and finally, above 40 inches of rainfall—in the temperature
region—completely disappears from the soil profile."

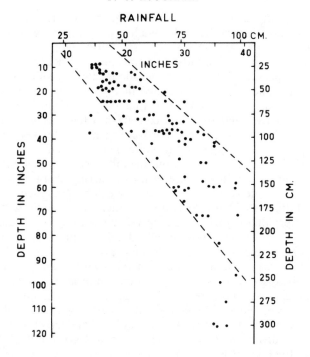

Figure 57.   The relation between rainfall and depth to the top of the carbonate accumulation in soils derived from loess (after Jenny and Leonard, 1934).

The relation of depth of the lime horizon to rainfall is shown in Fig. 57.

There appears to be general agreement among pedologists that mechanical eluviation of colloidal clay is a common phenomenon in soils, particularly those subjected to a high degree of leaching (Robinson, 1949). The development of soils with sandy textured A horizons and clayey B horizons is generally considered evidence of the effectiveness of this mechanism. Hallsworth (1963) studied the extent to which clay minerals can migrate through various artificial sand-clay mixtures by way of percolating solutions. He found that movement of the clay is restricted as the proportion increases and apparently ceases when the clay content exceeds 40% in the case of kaolinite and 20% in the case of montmorillonite. However, at these clay contents, significant migration occurs by way of cracks which develop as the soil dries out. The peculiar gilgai soils (see p. 134) are believed to have originated through migration of material, particularly clay, down cracks which developed during the drying of the soil.

The A and B horizons together comprise the *solum*.

The C horizon is a transitional zone of partly altered rock or saprolite, and is considered by pedologists as the parent material for the solum. Consequently, soil studies rarely extend below this zone. From the viewpoint of those primarily interested in rock weathering, this is unfortunate, for much valuable data are disregarded. The unaltered rock, or "bedrock" of the pedologists, is termed the D horizon. In well-developed profiles further differentiation of both the A and B horizons into distinctive subhorizons or layers is often possible. These layers are designated by subscripts, such as A_{00}, A_0, A_1, B_2, and so on, as indicated in Fig. 56. The thicknesses of the individual horizons vary considerably; thus the A horizon may range from an inch up to 4 feet, and the B horizon from a few inches to a probable maximum of 6 feet, whereas the C horizon is generally thicker and, in well-leached tropical soils, may extend to more than 100 feet.

In the early stages of soil formation, the influence of the parent rock type is most pronounced and the nature of its weathering characterizes the soil. Such soils are considered immature and may exhibit only incipient differentiation into horizons. Mature soils, on the other hand, are those which have evolved after prolonged periods of relatively constant climatic and drainage conditions and these may show profile development to an extraordinary degree. In these mature soils, the soil forming processes of mineral alteration, organic matter accumulation, and mineral translocation, and the soil-destroying processes, predominantly erosion, approach a state of equilibrium. Pedologists generally believe that, as a result of this balance, the influence of the parent rock is considerably suppressed, although probably never completely erased, and soils with similar characteristics are found on a wide array of diverse rock types. "Thus" wrote Carroll (1962), "a podsol in Australia, England, Russia, United States or elsewhere, has the same kind of profile, which is a consequence of it having been developed by similar soil-forming factors." These mature soils of world-wide distribution are known as the great soil groups.

CLASSIFICATION OF SOILS

Since the development of modern pedology, commencing with the Russian school toward the end of the past century, a considerable number of soil classifications have emerged with emphasis being placed at various times on differing characteristics such as color, texture, morphology, genesis, and so on. Even at present no universally accepted scheme is available but rather each country has tended to adopt or adapt that which has been considered most suitable to its needs. The commonly accepted category is the *great soil group* which, according to Bunting (1965), is "a group of soils having a wide distribution and a number of common fundamental internal characteristics."

Soils are generally classified as zonal, intrazonal, or azonal. The *zonal soils* are those which possess mature characteristics and which have developed over wide geographic areas under reasonably uniform climatic, drainage, and vegetational influences. The *intrazonal soils* also have mature profiles but differ from the prevailing zonal soils type because of the predominance of some local variation in the soil-forming factors such as parent material, relief and so on. The *azonal soils* generally bear little relation to the prevailing zonal type and are immature soils which have developed on recent sands (regosols), alluvium (alluvial soils), or bare rock (lithosols or skeletal soils).

The subdivision of the zonal soils into *pedocals* and *pedalfers* was due to Marbut (1928). These terms were partly derived from the Greek (*pedon*= ground) and partly from the Latin (*calcis*=lime, *alumen*=aluminum, *ferrum* =iron). The pedocals, therefore, are characterized by the presence of carbonate accumulations within the solum and are restricted to the more arid regions of the world. The pedalfers, on the other hand, are humid soils which lack carbonate accumulations but rather tend to develop sesquioxide (Fe_2O_3 $+Al_2O_3$) concentrations in parts of the profile. As a general rule, 25 inch rainfall marks the division between the arid pedocals and the humid pedalfers.

A diagrammatic representation of the distribution of the great soil groups according to climate is shown in Fig. 58.

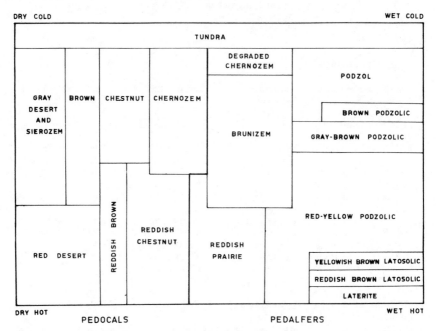

Figure 58. Diagrammatic representation of the distribution of some of the great soil groups according to climate after Millar, Turk, and Foth, 1958).

Hallsworth (1964) has suggested an interesting and rather simple classification of the major groups of mineral soils based on the extent of leaching to which the soils have been subjected. He has pointed out that "The more permanent features of the profile are those associated with the leaching pattern, with the nature and quantity of clay minerals left behind as a residue of weathering, with the size distribution of the mineral skeleton, whether the coarse particles of the soil are of primary or secondary nature, and the balance and distribution of the exchangeable cations." Hallsworth's proposals are summarized in Table 34. In the following brief descriptions of some of the great soil groups of the world, reference is made to the characteristic mineral constituents.

TABLE 34

Classification of the Major Groups of Mineral Soils by Extent of Leaching[a]

Group	Normal soils		Ground water soils
	Light-textured	Heavy-textured	
0. Soils showing no or only rudimentary differentiation	Lithosols Aeolian regosols Dune sands (Seif and barchan dunes, lunettes) Fluvial regosols	—	Alluvial soils
1. Soils effectively unleached, containing soluble salts, mainly sodium chloride	—	—	Solontchak
2. Slightly leached soils (2nd stage of leaching) dominated by Na$^+$, often containing gypsum	Seif dune soils Solonetz Solodized solonetz Solodic	Stony gilgai Serozem	Salty alkali (Takyr) Solontchak
3. Moderately leached soils A (3rd stage of leaching) dominated by Ca (Mg) and containing secondary carbonate	Red-brown earth Solonized soils	Chernozem Chestnut	Gleyed serozem Gleyed chernozem
4. Moderately leached soils B (4th stage of leaching) dominated by Ca(H) and without secondary carbonate	Brown earths Brown soils of light texture Brown limestone soils Sols lessive'	Prairie soils Brown earths Chocolate soils Terra rossa Rendzina	Meadow soils or Weisenboden Gleyed brown earths
5. Strongly leached soils (5th stage of leaching) dominated by H, Al, (Fe)	Podsols Gray-brown podsolic Red-yellow podsolic	Krasnozems Red earth	Gley podsol Laterite

[a] After Hallsworth (1964).

ZONAL SOILS

Tundra Soils

The tundra soils occur in the arctic to subarctic regions where the typical vegetation is moss and similar lowly forms. These soils have a permanently frozen zone below the surface, the permafrost, and as a result drainage is impeded. The accumulation of peat is favored by the development of marshy conditions during the warmer months following the thaw. Because of the low temperatures and poor drainage, chemical reactions are retarded and the inorganic components of the soil consist mainly of the products of physical rather than chemical weathering. Consequently, there is generally little new development of clay minerals and those which do form are of the chlorite, montmorillonite and illite groups, most commonly mixed layers of these minerals. The peaty surface layers often have a hummocky appearance due to partial erosion and characteristically these are underlain by a compact, bluish-gray water-logged subsoil containing some clay. The waterlogged subsoil is in a reduced state and referred to as the glei (gley) horizon. The pH values vary from mildly alkaline to strongly acid in the peat layers.

Desert Soils

The desert soils are variously gray, brown, or red depending on the temperatures while the textures may range from heavy clay to light loams and friable, loose sands. Because of the low humidity in these regions, chemical weathering is retarded and the scant vegetation is generally insufficient to prevent excessive wind erosion nor permit accumulations of organic matter. Consequently, desert soils tend to be immature and, because of deflation, often consist of residual concentrations of rock fragments. These are the skeletal soils of the stony deserts. Owing to ineffective leaching, lime and salts such as gypsum are frequently present and the soils, for the most part, are alkaline. However, acid desert soils are known (Teakle, 1936; Crocker, 1956). The paucity of organic matter, particularly in the warmer deserts, and the absence of a permanently saturated zone favor oxidation. Consequently, iron released from the parent rock persists in the ferric state and, as such, tends to coat grains and bare rock alike rendering the soils reddish and imparting a "varnished" appearance to exposed rock and pebble surfaces. Because of the lack of maturity, desert soils tend to reflect the parent rocks from which they have been derived and, often within relatively small areas, a considerable variety of soils exists. Much of the soil is composed of unaltered to partly altered parent minerals while the retention of the alkalies and alkaline earths favors the development of illite and montmorillonite or mixed layers of these minerals as the characteristic secondary products. Alexander et al. (1939) and Hseung and Jackson (1952) have recorded the relative abundance of illite and expandable lattice minerals in desert soils from California and China, respectively. However, according to Buol (1964) kaolin minerals may predominate in some arid and semiarid soils. Hosking et al. (1957) also found kaolinite more abundant than illite in a semidesert soil developed on granite in South Australia.

Chestnut Soils

The chestnut soils are not particularly widespread but in the northern hemisphere they develop on the arid side of the more abundant chernozems, principally in the 12–16 inch rainfall belt. Chestnut soils have been described from the southern hemisphere, notably South Africa and South America. The profile consists of a gray-brown A horizon in which the A_1 generally has a platey structure and contains up to 6% organic matter, while the A_2 is a lighter brown owing to the presence of lime, often in association with gypsum. The lime extends down into the B horizon which is light-colored and has a heavier texture due to illuviated clay. Little data are available on the mineralogy of the chestnut soils but from the work of Hseung and Jackson (1952) it would appear to be intermediate between those of the desert soils and the chernozems.

Chernozems or Black Earths

The chernozems are the fertile grassland soils of the 12–25 inch rainfall areas occurring mainly in the temperate regions although tropical and subtropical chernozems are known. These soils are particularly well developed in the Ukraine, where the name was originally applied by Dokuchaiev, and in North America they form a wide belt extending from Saskatchewan and Alberta in the north through the Dakotas and Nebraska down into Oklahoma and Texas. In Australia, chernozems are virtually restricted to the eastern states, particularly northern New South Wales and southern Queensland, where the parent rock is basalt of Tertiary age.

The chernozem profile (Fig. 59) is composed of a thick (1–4 feet) black and only slightly differentiated A horizon underlain by a light yellowish to brownish B horizon. The A horizon has a granular structure and contains organic matter, generally to the extent of 8–10%, although in places the content may be considerably greater. Lime is invariably present in the B horizon either disseminated throughout, in which case it may extend into both the overyling and underlying zones, or, more commonly, concentrated in a well-defined layer near the base of the horizon. The top of the profile is slightly acid (pH > 6) but the pH increases downward, becoming alkaline near the base of the A horizon. Despite the strongly reducing environment created by the abundant organic matter, leaching is apparently insufficient to bring about translocation of iron from the A to the B horizon.

The retention of the alkaline earths favors formation of minerals of the montmorillonite-illite groups although the increased leaching and acid conditions near the top of the profile may lead to the development of a little kaolinite in that zone. Ferguson (1954) and Hosking et al. (1957) investigated the mineralogy of chernozems derived from basalt in eastern Australia and found the only clay minerals present are members of the montmorillonite group. Montmorillonite also proved to be the dominant clay mineral in till-derived chernozems in North Dakota (Redmond and Whiteside, 1967). However, Alexander et al. (1939) analyzed the B_2 horizon of chernozems formed on till in South Dakota and found the montmorillonite content to be somewhat subordinate to that of illite and kaolinite. Hseung and Jackson (1952) also found illite the most abundant clay mineral in a chernozem developed on loess in China.

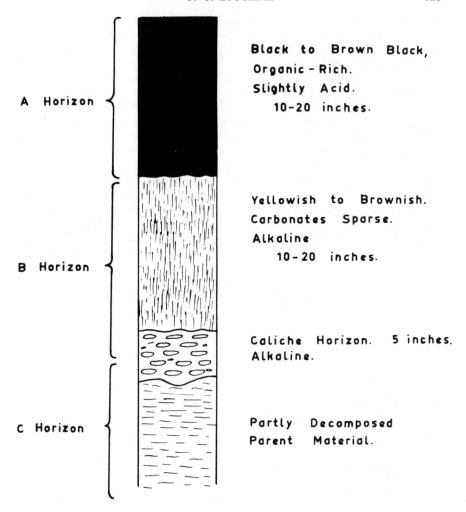

Figure 59.   A chernozem profile from North Dakota.

Red-Brown Earths

In the 12–25 inch rainfall belts of the lower latitudes, organic matter tends to be oxidized more rapidly than it accumulates and, as a result, the cherno-zems and chestnut soils of the higher latitudes are replaced by red-brown earths. In Australia, red-brown earths are widespread and in the southern part of the continent they form the basis for the wheat-growing industry.

These soils have a brown to red-brown A horizon which may contain a little organic matter near the top (A_1) underlain by a deeper red B horizon which has a compact structure. The A horizon measures about 12 inches while the B is a little thicker. Lime occurs near the base of the B horizon and may extend into the C. The A horizon is slightly acid but the pH increases with depth and below the sharply defined boundary between the A and B horizons, the soil is alkaline.

Radoslovich (1958) made a study of the clay minerals in the B horizon of a number of Australian red-brown earths derived from a variety of parent materials ranging from alluvium to crystalline igneous rocks. He found the parent material to be the most influential factor controlling the development of secondary clay minerals (Table 35). Alluvium and similar materials yield soils with illite in excess of kaolinite whereas igneous rocks, including both acid and basic, give rise to soils in which kaolinite is the dominant constituent.

Hosking et al. (1957) also examined the clay minerals in red-brown earths derived from a hornblende granite, a basalt and alluvium composed of shale and slate fragments (Table 35). The soils developed from the granite contain

TABLE 35

Clay Minerals in Some Australian Red-Brown Earths[a]

| No. | Parent material | Horizon | Clay minerals[b] | | | | |
			Il	K	Mt	M.L.	Others
1	Alluvium	B_1	v.c.	c	—	—	Chlr?
2	Proluvium	B_1	a	v.c.	—	c	—
3	Shales	B_1	a	v.c.	—	v.c.	—
4	Solontchakous sediments	B	v.c.	c	c	v.c.	—
5	Coal measure sediments	B_1	v.c.	v.c.	—	—	Chlr or Vm?
6	Alluvium	—	a	v.c.	c	—	—
7	Basaltic sediments	B	—	a	—	c	—
8	Alluvium	B	—	v.a.	?	c	—
9	Granodiorite	B	c	a	—	c	—
10	Granite	B	—	a	—	c	—
11	Metasediments	B_2	—	a	—	v.c.	—
12	Basic alluvium	B	v.c.	v.c.	v.c.	c	—
13	Basic alluvium	BC	c	a	—	?	—
14	Granite	BC	—	a	v.c.	c	Chlr or Vm (c)
15	Dolerite colluvium	B	—	—	v.c.	c	—
16	Dolerite	B	—	c	c	v.c.	—
17	Granite	BC	c	a	c	—	—
18	Basalt	B	r	c	a	—	—
19	Alluvium	A	v.c.	v.c.	—	—	—

[a] Nos. 1–16 taken from Radoslovich (1958); Nos. 17–19 taken from Hosking et al. (1957).
[b] Il=illite; K=kaolinite and (or) halloysite; Mt=montmorillonite and (or) nontronite; M.L.=mixed-layered clay minerals; Chlr=chlorite; Vm=vermiculite. v.a.=very abundant (>80%); a=abundant (50–80%); v.c.=very common (20–50%); c=common (10–20%); r=rare (<10%).

abundant halloysite with a little nontronite and illite whereas the basaltic soil is characterized by a predominance of nontronite in considerable excess of kaolinite and illite. Halloysite and illite occur in approximately equal amounts in the soil derived from the alluvium.

Prairie Soils and Brunizems

The prairie soils and brunizems form in the intermediate zone between the grassland and steppes of the less-humid chernozems and the leached podsolic soils of the forest areas. As such, they are transitional, having affinities with both the pedocals and pedalfers. The profile resembles that of the chernozems with a thick, dark, organic-rich A horizon underlain by a yellow to light-brown B horizon which is enriched in clay derived from above. However, unlike the chernozems, lime is absent from the solum, although, where the parent rock is calcareous, it may appear in the C horizon. Whereas leaching has been sufficient to remove lime from the solum there is little movement of the sesquioxides and the ortstein layer of the podsols is absent. The A horizon has a granular texture and is moderately acid but the pH rises with depth, becoming neutral toward the base of the B horizon.

In North America, the prairie soils attain their ultimate development on the broad till and loess deposits of Illinois and adjacent states. Figure 60 is a typical profile of a prairie soil from Illinois and Table 36 shows the mineralogical analyses of a similar section, after Beavers *et al.* (1954).

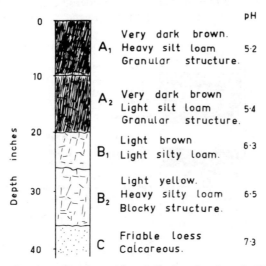

Figure 60.   A profile of a prairie soil developed on loess in Illinois.

Podsols and Podsolic Soils

The term podsol covers a wide range of leached soils which generally exhibit profile development to an extraordinary degree. Although these soils

TABLE 36

Analyses of the Elliott Silt Loam (Brunizem) from Illinois[a]

Horizon	Depth (in.)	Mineralogy (%)			pH	Organic carbon (%)
		Montmoril-lonite	Illite	Chlorite		
A	0–12	5	80	15	5.9	4.15
B	12–24	10	65	25	7.1	0.68
C	24–43	0	65	35	8.0	0.47

[a] After Beavers *et al.* (1954).

are more prominent in the cool, humid, forested areas of the world, they are also relatively common in the subtropics and may occur in the tropics. In a mature podsol profile (Fig. 61) the A horizon is differentiated into an A_{00} of forest leaf litter, an A_0 of raw humus, an A_1 of dark organic matter, and an A_2 of light gray to off-white leached sand which has been depleted of humus and sesquioxides. The B horizon is composed of a B_1 layer containing dark-brown, compact humus, underlain by a deep red-brown, indurated B_2 of sesquioxide accumulation which forms a hardpan known as the *ortstein layer*. This grades into a B_3 composed of iron-stained, permeable, sandy material. All layers may not be present but those considered characteristic, and hence essential to the identification of a podsol, are the A_1, A_2, and B_2.

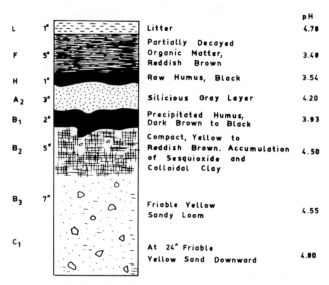

Figure 61.   A well-developed podsol profile (after Lyon and Buckman, 1943).

Because of the intense leaching and relative abundance of organic matter, podsolic soils are strongly acid and this considerably influences the stability of the inorganic constituents. Not only are the alkalies, alkaline earth, and ferrous iron removed but also there is abundant evidence attesting to the translocation of alumina and ferric iron from the A to the B horizon. Whether this is achieved through solution or by migration of colloidal material is not known but certainly the pH values are such as to render aluminum soluble, while reduction of the iron to the soluble ferrous state would be expected in the organic-rich A horizon. The actual position of the B_2 layer in the podsol profile appears to be governed by the upper limit of the water table. Consequently, it is possible that at least some of the iron in the B_2 has been derived from the underlying C horizon and has been deposited in much the same manner as the concretionary zone of the laterites. Although podsols may form on a variety of parent materials their development is favored by permeable rocks low in bases, particularly friable and unconsolidated sands. Robinson (1949) believed the distribution of podsols to be markedly influenced by geology and contended, "if the whole of England were a region of light sands, such as the Bagshot Beds, it would be represented on a soil map as an area of podsols."

The influence of moderate to high rainfall and unimpeded drainage is reflected in the mineralogy of podsolic soils. Salts and carbonates are absent and the parent material tends to be converted to assemblages of sesquioxide-rich minerals, such as hematite, kaolinite, and possibly gibbsite. Bryant and Dixon (1964) have described the mineralogy of an interesting podsolic soil formed from quartz-mica schist in Alabama. Mica and interlayered chlorite-vermiculite are the most abundant minerals in the clay fraction of the A horizon, whereas in the B horizon kaolinite and gibbsite predominate. The sand and coarse silt fractions, however, showed an increase in mica and a decrease in quartz contents with depth. They attributed the abundance of kaolinite and gibbsite in the B horizon to eluviation-illuviation processes.

On the other hand, Alexander *et al.* (1939) studied the mineralogy of the fine fraction of a number of podsolic soils developed on granite, limestone, and till in eastern United States and found illite to be major constituent (Table 37).

Lateritic Soils including the Krasnozems (Latosols)

The term *laterite* (Latin *later*=brick) was originally introduced by Buchanan-Hamilton in 1807 to describe earthy and vesicular, ferruginous crusts which were being cut into bricks for building purposes by the inhabitants of south-central India. Over the intervening period, the term has been given various meanings leading to much confusion. However, according to Jenny (1950), "In recent years . . . the concept of laterite has tended to be restricted to specific soil strata rich in sesquioxides, possibly formed under the influence of ground-water conditions."

TABLE 37

Mineralogical Analyses of the Fine Fractions (0.3 μ) of Some Podsolic Soils[a]

Soil series—location	Great soil group	Parent rock	Horizon	Mineralogy[b] (%)					Clay, -2 μ (%)
				K	Hm	Mt	Fe	Qtz	
Cecil, N. C.	Red podsolic	Granite	C	80	—	—	15	—	44
Grenville, Ga.	Red podsolic	Granite	B₂	90	—	—	10	—	42
Decatur, Ala.	Red podosolic	Limestone	B₂	70	10	—	13	—	49
Dewey, Ala.	Red podsolic	Limestone	B₂	70	10	—	13	—	36
Fullerton, Ala.	Yellow podsolic	Limestone	B₂	60	20	—	8	10	49
Frederick, Va.	Gray-brown podsolic	Limestone	B₂	70	20	—	10	—	70
Hagerstown, Pa.	Gray-brown podsolic	Limestone	B₂	70	20	—	9	—	55
Hagerstown, Md.	Gray-brown podsolic	Limestone	B₂	50	40	—	8	—	70
Hagerstown, Mo.	Gray-brown podsolic	Limestone	B₂	40	40	—	7	5	39
Chester, Md.	Gray-brown podsolic	Granite	A₂	50	30	—	11	—	20
Chester, Md.	Gray-brown podsolic	Granite	B₂	50	30	—	15	—	25
Chester, Md.	Gray-brown podsolic	Granite	C	50	30	—	18	—	17
Manor, Va.	Gray-brown podsolic	Granite	B₂	50	40	—	10	—	24
Miami, Indiana	Gray-brown podsolic	Till	B₂	10	80	—	7	—	51

[a] After Alexander *et al.* (1939); used by permission of The Williams and Wilkins Co.
[b] K=kaolinite; Hm=hydromuscovite (illite); Mt=montmorillonite; Fe=iron oxides; Qtz=quartz.

Although lateritic soils are common in the tropics, particularly in well-drained areas subject to a savanna-type rainfall, it should be emphasized that not all tropical soils are lateritic. The essential requirements for their formation appear to be a high rainfall, intense leaching, and a strongly oxidizing environment. Under these conditions organic matter accumulation is inhibited while the leaching solutions tend to flush out all potentially mobile constituents leaving concentrations of alumina, titania, and ferric oxide. Silica may persist to the extent necessary to combine with the available alumina in the formation of kaolinite but frequently it is deficient and the excess alumina is present as gibbsite or, more rarely, boehmite. Where present, gibbsite generally attains maximum development in the more highly leached surface horizons. However, Glenn and Nash (1964) have described two reddish brown lateritic soils formed on loose sediments in southern Mississippi, in which the gibbsite contents increase appreciably with depth. These authors believe that the prevailing pH and the presence of expandable lattice clay minerals have had a controlling influence on the distribution of gibbsite within the soils. At the surface where the pH values are unusually low (< 4.8), alumina tends to enter the lattice of the expandable clay minerals and converts the latter to a chloritelike mineral. This is the "antigibbsite effect" described by Jackson (1963). As the pH increases with depth, the alumina becomes progressively more negative and cannot enter the clay mineral structures. Consequently it crystallizes as gibbsite. Generally, however, the surface horizons of laterites are only mildly acid (Table 38) despite the absence of bases and the intense leaching. This is attributable to the low cation exchange capacities of the residual crystalline phases.

TABLE 38

Mineralogy and Characteristics of a Laterite Profile in Southwestern Australia[a]

Zone	Depth	Description	pH	Mineralogy,[b] 2 μ
Ferruginous	0–5 feet	Yellowish-brown hard ironstone becoming slightly softer toward the base	6.5	Kaolinite, m Gibbsite, m Quartz, L Hematite, l
			5.2	Kaolinite, m Gibbsite, M Quartz, l Hematite, l
Pallid	5–23 feet	White and slightly pinkish clay with dark red patches.	5.3	Kaolinite, M
		Quartz grains are bleached in the white clay, stained with iron oxide in the red.	5.1	Kaolinite, M Montmorillonite, l Inhibited vermiculite, l Quartz, l
		Occasional mica flakes throughout. Continuing	4.9	Kaolinite, M Illite, L Montmorillonite, l Inhibited vermiculite, l
			6.0	Quartz, l
Transitional zone		Pale brown and rusty mottled weathered gneiss		

[a] After Mulcahy (1960).
[b] M=much; m=moderate; L=little; l=very little.

In the lateritic red earths the A horizon consists of a reddish loam which may contain a little organic matter, while the thick B horizon is composed of cavernous, vesicular or pisolitic iron oxide with kaolinite or bauxite minerals or both. According to Sherman and Kanehiro (1954), much of the concretionary iron oxide in the lateritic soils of the Hawaiian Islands has magnetic properties and is probably maghemite rather than hematite. A sharply defined boundary separates the B horizon from the underlying mottled zone which is kaolinite-rich and may extend to a depth of 20 feet or more. This zone is believed to represent the upper and lower limits of the

TABLE 39

Mineralogy of Some Latosols (Krasnozems) from the Hawaiian Islands[a]

Soil	Rain (in.)	Horizon depth (in.)	Percent of soil	Mineralogy[b] (%)													
				Qtz	SiO$_2$	M	Int	Vm	Mt	K	All	Gb	Ma	Gt	Hm	An	Total
Low humic	15–45	A, 0–8	92.2	—	—	—	—	—	9	54	11	3	0	0	17	3	97
		B, 8–18	82.4	—	—	—	—	—	6	56	11	3	0	0	19	2	97
Humic	45–150	A, 0–8	76.8	5	—	5	12	0	8	16	0	16	0	8	24	7	101
		B, 8–15	91.5	1	—	1	8	1	10	19	0	16	0	20	19	4	99
Hydrol humic	150–500	A, —	67.7	2	5	7	0	0	0	0	16	27	18	12	4	7	98
		B$_1$, —	85.5	2	4	5	0	0	0	0	26	26	11	18	4	8	104
		B$_2$, —	87.5	2	4	10	0	0	0	0	15	25	6	34	0	7	103

[a] After Tamura et al. (1953). Reproduced from the *Proc. Soil Sci. Soc. Am.*, **17**, 345, by permission of the publishers.

[b] Qtz = Quartz; SiO$_2$ = silica; M = mica; Int = mixed-layered clay minerals; Vm = vermiculite; Mt = Montmorillonite; K = kaolin; All = allophane; Gb = gibbsite; Ma = magnetite; Gt = goethite; Hm = hematite; An = anatase.

fluctuating water table. The permanently saturated pallid zone below the mottled zone consists of partly decomposed parent rock which has lost much of its original iron content through solution and upward migration toward the oxidized parts of the profile.

The krasnozems, also known as latosols and red loams, are normally developed on parent rocks relatively rich in iron such as basalts, dolerites, and ferruginous sediments. These soils have a very deep profile but, apart from a small visible accumulation of organic matter at the surface, there is little differentiation down to the parent rock. They are strongly acid with a friable texture and, as Hallsworth (1951) has pointed out, despite the apparent red color, they contain a surprising amount of organic matter. Krasnozems occur in the moist subtropics and their unusual features have been attributed to the flocculating effect of hydrated ferric oxide on the clay, which is essentially kaolinitic.

Tamura et al. (1953) have demonstrated the effect of increasing rainfall on the mineralogy of a series of latosols in the Hawaiian Islands (Table 39). The low humic latosols developed in the 15 to 45 inch rainfall belt are essentially kaolinitic, but with greater rainfall gibbsite and goethite become important constituents and in the hydrol humic soils where the annual rainfall exceeds 150 inches per year these minerals predominate. The mineralogy of a number of krasnozems developed on basalt in northern Queensland has been described by Simonett and Bauleke (1963). Despite a variation in the mean annual rainfall, the mineral content of the clay fractions of these soils is remarkably constant (Table 40). Kaolinite is the dominant constituent in all whereas gibbsite shows only sporadic development.

Intrazonal Soils

The Halomorphic Soils

These are alkaline soils of the arid zone, characterized by the presence of excess sodium salts. They include the *solontchaks* or "white alkali soils" (so named because of the white efflorescence which develops during dry periods) and the *solonets* or "black alkali soils." In the solontchaks the sodium occurs as the chloride, generally in association with the sulfate and calcium salts, whereas in the solonets sodium carbonate predominates. Both soil types tend to develop in depressions where the ground waters are at shallow depth, in areas of normal occurrence of chernozems, chestnut soils, and red-brown earths. The source of the salts may have been the parent rock but mostly it is the residue of preexisting lakes since dried up, or, as appears to be the case in many parts of Australia, it is of cyclic origin. The solontchaks are very variable but frequently show little profile development. On the other hand, the more humid solonets have a shallow A horizon which in the higher latitudes, at least, contains appreciable organic matter, and a markedly prismatic B horizon. The latter feature is generally considered diagnostic for a solonets and is attributed to the deflocculated nature of the clay.

TABLE 40

Mineral Content of Clay Fraction of Krasnozem Soils Developed on Basalt, North Queensland[a]

Mean ann. rainfall (in.)	Depth (in.)	Mineralogy[b] (%) Hm	Gt	Ma	K	Gb	Qtz
36	0–6	9	—	6	77	Trace	3
	12–18	6	—	7	74	1	2
	18–30	8	—	4	79	3	3
36	0–4	—	7	6	79	—	3
	4–20	—	7	8	76	—	3
	20–39	—	7	9	78	—	1
	42	—	7	8	78	—	3
40	0–4	8	—	7	72	1	—
	4–12	—	8	7	74	1	—
	12–36	—	7	10	69	2	—
	36–48	—	8	8	73	—	1
	48	—	8	8	75	—	1
52	0–3	8	—	6	66	13	—
	12–24	10	—	6	66	12	—
	36–48	10	—	7	72	7	—
55	12–24	9	—	10	71	2	—
	24–36	10	—	6	74	6	—
64	0–12	11	—	9	70	1	—
	12–60	8	—	7	78	2	—
75	0–8	6	—	11	61	6	—
	8–20	7	—	8	66	3	—
	20	6	—	15	Nd.	—	—
76	0–6	—	8	10	62	Trace	—
	6–12	—	8	11	68	1	—
	18–30	—	6	12	70	—	3
	30–42	—	8	11	69	—	3
	42–54	—	10	11	70	—	
147	0–4	12	—	8	52	19	—
	4–30	11	—	7	60	13	—
	36–40	6	—	7	68	9	—
	72	5	—	8	67	7	—

[a] After Simonett and Bauleke (1963). (Reproduced from *Proc. Soil Scr. Soc. Am.* **27**, 209 by permission of the publishers.)
[b] Hm=hematite; Gt=goethite; Ma=magnetite; K=kaolinite; Gb=gibbsite; Qtz=quartz

The Hydromorphic Soils

The hydromorphic soils, including the weisenbodens (meadow soils), ground water podsols, ground water laterites, planosols, and the various peats, are those in which the water table is sufficiently close to the surface to influence profile development. They are most commonly found in depressions and low-lying areas but may also occur in high terrains where drainage is impeded by the presence of an impervious subsoil. In either case rainfall, or rather water influx, must exceed evaporation. Accumulation of organic matter is greatly facilitated by the saturated conditions and most hydromorphic soils possess a dark, peaty surface zone. Underlying this is generally a bluish-gray to mottled zone which owes its appearance to reduction of much of the iron to the ferrous state during protracted periods of waterlogging. This is termed the *glei* (gley) horizon. The downward migration of clay is often quite marked, particularly in the planosols, where it forms an indurated layer of low permeability known as a *hardpan*. The hydromorphic soils range from very acid with pH values below 4 in the case of some of the better-drained high moor peats, to neutral or even mildly alkaline in the low-lying peats where bases have probably been introduced by way of drainage waters from nearby hills.

The Calcimorphic Soils

Where calcareous-rich rocks such as limestones form the parent material, the derived soils tend to be of two distinctive groups, the *terra rossas* and *rendzinas*. Terra rossa, as the name implies, are red to red-brown soils which show little differentiation into A and B horizons. Generally they are moderately acid and consist of abundant clay and iron oxide with a little organic matter which may darken the color of the surface zone. The boundary between the solum and the parent rock is often remarkably sharp. It is generally believed that the inorganic material comprising these soils represents the residue from solution of the limestone. The rendzinas, on the other hand, are dark-colored, gray to black, alkaline soils which contain appreciable amounts of organic matter and free calcium carbonate. Like the terra rossas, they show little horizon differentiation. The relationship between the terra rossas and the rendzinas is not fully understood since they frequently occur together over a wide range of climatic conditions. According to Stephens (1962), in Australian occurrences of these soils, the terra rossas tend to be associated with the harder limestones and the rendzinas with the softer varieties. However, Stace (1956) has described a typical profile of each soil type from South Australia, both of which have apparently developed on soft calcareous parent materials (Table 41).

The composition of the calcimorphic soils varies greatly and probably is strongly influenced by the nature of the parent material. In the study of the clay fractions of a number of terra rossas and rendzinas from southeastern Australia, Norrish and Rogers (1956) found little difference in the mineralogy of the two soil groups. Illite and kaolinite are the dominant constituents of both. Ravikovitch *et al.* (1960), on the other hand, have shown that two distinctive mineral associations occur in the calcimorphic soils of Israel (Table 42). Terra rossas are characterized by abundant kaolinite, lesser

TABLE 41

Morphological Characteristics of a Typical Terra Rossa and a Typical Rendzina from South Australia[a]

Terra Rossa
0.4 in. red-brown soft loam with granular structure.
4–11 in. red soft loam—sandy clay loam with nutty structure.
11–18 in. red soft sandy clay loam—sandy clay with nutty structure. This is sharply separated from hard calcareous material overlying soft calcareous material.

Rendzina
0–4 in. near black friable clay with granular structure. Few carbonate granules.
4–8 in. black friable clay with coarse granular structure. Few carbonate granules. This is sharply separated from
Irregular band, approximately 1½ inches thick, of grayish white, calcareous material. This passes into
Pinkish-yellow, soft calcareous material.

[a] After Stace (1956).

TABLE 42

Mineralogy of the Colloids in Some Calcimorphic Soils of Israel[a]

Locality	Origin	Mineralogy; order of abundance
	Terra Rossa	
Metallua	Hard limestone	Illite, kaolin, quartz
Eilon	Hard limestone and dolomite	Kaolin, hematite, quartz, montmorillonite
Ramot Naptholi	Hard limestone	Kaolin, illite, quartz
Safad	Hard limestone	Kaolin, illite, quartz
Ma'ala Haha mischa	Hard limestone	Kaolin, montmorillonite, illite, quartz, hematite
	Mountain Rendzina	
Kefar Uriya	Marl	Montmorillonite, calcite, illite, kaolin, quartz
	Valley Rendzina	
Messilot	Lissan Marl	Montmorillonite, calcite, attapulgite, illite, kaolin
Beit Shean	Lissan Marl	Montmorillonite, calcite, attapulgite
Ashdot Ya'agov	Lissan Marl	Montmorillonite, illite, calcite, kaolin, quartz

[a] After Ravikovitch *et al.* (1960).

amounts of illite, and a general paucity of montmorillonite, whereas montmorillonite is the major constituent of the rendzinas and is often associated with appreciable quantities of palygorskite (attapulgite).

Gilgai Formation

In many areas in Australia the soils are characterized by a hummocky appearance which is commonly termed *gilgai* or *crabhole* country. The mounds are approximately circular in outline and of the order of 3 feet across while the intervening depressions are generally slightly wider. The difference in elevation between mound and depression ranges from an inch or so up to several feet. Although the surface soils of both mound and depression are dark gray in color, they contrast markedly with each other in texture and composition. Whereas the soils of the mounds are calcareous and crumble upon drying, those in the depressions are noncalcareous and dry to hard clods. Moreover, the mounds generally carry annual plants while perennials predominate in the depressions.

According to Leeper (1964), two conditions are necessary for the formation of gilgai soils, (a) a climate of alternating wet and dry seasons and (b) the presence of abundant swelling clay minerals in the soil (Fig. 62). The heavy-textured soils upon drying crack down into B_2 horizon and clay from the surface is blown into these cracks. Upon rewetting, the clay filling the cracks swells, thrusting the soil upward and outward. The upthrust portion is then prone to wind erosion during the next dry spell. Many cycles of

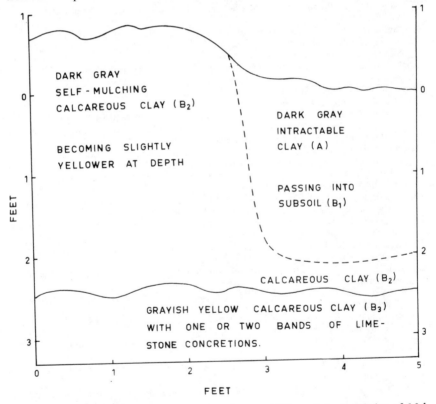

Figure 62.   Section of a gilgai complex (after Leeper, 1964; used by permission of Melbourne University Press). A_1, B_1, B_2, and and B_3 refer to horizons.

wetting and drying result in exposure of the B_2 horizon at the surface. However, Rix and Hutton (1951) have described gilgai formation in sandy soils from the Mt. Lofty Ranges in South Australia and have shed some doubt on this mechanism.

Gilgai formation has also been recorded from Kenya in Africa (Stephen et al., 1956), where development has taken place in montmorillonitic, tropical black earths derived from alkaline lavas.

REFERENCES

Alexander, L. T., S. B. Hendricks, and R. A. Nelson (1939), Minerals present in soil colloids. II. Estimates in some representative soils, Soil Sci. **48**, 273–280.

Beavers, A. H., W. D. Johns R. E. Grim, and R. T. Odell (1954), Clay minerals in some Illinois soils developed from loess and till under grass vegetation, Proc. Nat. Conf. Clays and Clay Minerals **3**, 356–372.

Bryant, J. P., and J. B. Dixon (1964), Clay mineralogy and weathering of a red-yellow podsolic soil from quartz mica schist in the Alabama piedmont, Proc. Nat. Conf. Clays and Clay Minerals **12**, 509–522.

Buol, S. W. (1965), Present soil forming factors and processes in arid and semi arid regions, Soil Sci. **99**, 45–49.

Bunting, B. T. (1965), The Geography of Soil. Hutchinson Univ. Press, London.

Carroll, D. (1962), Sedimentary Petrography. Vol. II. Principles and Applications, 4th ed. Milner, Allen and Unwin, London.

Crocker, R. L. (1956), The acid soils of the San Diego Mesa, California, J. Soil. Sci. **7**, 242–248.

Ferguson, J. A. (1954), Transformations of clay minerals in black earths and red loams of basaltic origin, Australian J. Agr. Res. **5**, 98–108.

Glenn, R. C., and V. E. Nash (1964), Weathering relationships between gibbsite, kaolinite, chlorite and expansible layer silicates in selected soils from the lower Mississippi coastal plain, Proc. Nat. Conf. Clays and Clay Minerals, **12**, 529–548.

Hallsworth, E. G. (1951), An interpretation of soil formations as found on basalts in the Richmond-Tweed region of New South Wales, Australian J. Agr. Res. **2**, 411–428.

Hallsworth, E. G. (1963), An examination of some factors affecting the movement of clay in an artificial soil, Soil Sci. **14**, 360–371.

Hallsworth, E. G. (1964), The relationship between experimental pedology and soil classification. Exptl. Pedol., 11th East. School, Nottingham, pp. 354–372.

Hosking, J. S., M. A. Nielson, and A. R. Carthew (1957), A study of clay mineralogy and particle size, Australian J. Agr. Res. **8**, 45–74.

Hseung, Y., and M. L. Jackson (1952), Mineral composition of the clay fraction of some main soil groups of China, Proc. Soil Soc. Am. **16**, 294–297.

Jackson, M. L. (1963), Interlayering of expansible layer silicates in soils by chemical weathering, Proc. Nat. Conf. Clays and Clay Minerals **11**, 29–46.

Jenny, H. (1950), Origin of soils. Applied Sedimentation (P. D. Trask, ed.), pp. 41–61. Wiley, New York.

Jenny, H., and C. D. Leonard (1934), Functional relationships between soil properties and rainfall, Soil Sci. **38**, 363–381.

Leeper, G. W. (1964), Introduction to Soil Science, 4th ed. Melbourne Univ. Press, Melbourne.

Lyon, T. L., and H. O. Buckman, The Nature and Properties of Soils, 3rd ed. Macmillan, New York.

Marbut, C. F. (1928), A scheme for soil classification, Proc. 1st. Intern. Congr. Soil Sci. **4**, 1–31.

Millar, C. E., L. M. Turk, and H. D. Foth (1958), Fundamentals of Soil Science, 3rd ed. Wiley, New York.

Mulcahy, M. J. (1960), Laterites and lateritic soils in southwestern Australia, *J. Soil Sci.* **11**, 206–225.

Norrish, K., and L. E. Rogers, (1956), The mineralogy of some terra rossas and rendzinas of South Australia, *J. Soil Sci.* **7**, 294–301.

Radoslovich, E. V. (1958), Clay mineralogy of some Australian red-brown earths, *J. Soi Sci.* **9**, 242–250.

Ramann, E. (1911), *Bodenkunde*. Springer, Berlin.

Ravikovitch, S., F. Pines, and M. Ben-yair (1960), Composition of colloids in soils of Israel, *J. Soil Sci.* **11**, 82–91.

Redmond, C. E., and E. P. Whiteside (1967), Some till-derived chernozem soils in eastern North Dakota: II. Mineralogy, micromorphology and development, *Proc. Soil Sci. Soc. Am.* **31**, 100–107.

Rix, C. E., and J. T. Hutton (1951), Gilgai soils, *Australian J. Sci.* **14**, 92–93.

Robinson, G. W. (1949), *Soils, Their Origin, Constitution and Classification*, 3rd ed. Thos. Murby, London.

Sherman, G. D., and Y. Kanehiro (1954), Origin and development of ferruginous concretions in the Hawaiian latosols, *Soil Sci.* **77**, 1–8.

Simonett, D. S., and M. P. Bauleke (1963), Mineralogy of soils on basalts in North Queensland, *Proc. Soil Sci. Soc. Am.* **27**, 205–212.

Stace, H. C. T. (1956), Chemical characteristics of terra rossas and rendzinas in South Australia, *J. Soil Sci.* **7**, 280–293.

Stephen, I., E. Bellis, and A. Muir (1956), Gilgai phenomena in tropical black clays of Kenya, *J. Soil Sci.* **7**, 280–293.

Stephens, C. G. (1962), *A Manual of Australian Soils*, 3rd ed. C.S.I.R.O., Melbourne.

Tamura, T., M. L. Jackson, and G. D. Sherman (1953), Mineral content of low humic, humic and hydrol humic latosols of Hawaiia, *Proc. Soil Sci. Soc. Am.* **17**, 343–346.

Teakle, L. J. H. (1936), The red and brown hard-pan soils of the acacia semi-desert scrub of Western Australia, *J. Dept. Agric. W. Australia* **13**, 480–489.

Glossary

ABRASION pH. The equilibrium pH value attained on grinding a specific mineral in distilled water.

ACTIVITY. The effective concentration of an ion in solution.

AMORPHOUS. Without a crystalline structure such as glass. See X-AMORPHOUS.

AMPHIBOLITE. A recrystallized metamorphic rock containing abundant amphiboles generally in association with plagioclase feldspar.

ANGSTROM UNIT (Å). 10^{-8} cm.

ANION. A negatively charged atom.

APLITE. A fine-grained dike rock composed of alkali feldspar and quartz and generally occurring within a granite mass.

ARID CLIMATE. One in which evaporation exceeds precipitation.

AUTHIGENESIS. The process involving the formation in situ of new minerals within a sediment.

BASAL SPACING. The unit distance separating two parallel planes in a crystal such that the planes intersect only the vertical or c crystallographic axis.

BASALT. A relatively fine-grained, basic, volcanic rock composed essentially of plagioclase feldspar and a pyroxene with or without olivine.

BAUXITE. A weathered residue composed essentially of the aluminum hydroxide minerals: gibbsite, boehmite and diaspore.

BENTONITE. A clay composed essentially of montmorillonite derived from the decomposition of volcanic ash.

CALICHE. A concentration of the carbonates, calcite and dolomite, in the lower part of the solum of soils of arid and semi-arid regions.

CAPILLARY ACTION. The upward movement of water through small interstices in a soil or rock. Movement is brought about by the molecular attraction of the water to the surfaces of the interstices.

CATION. A positively charged atom.

CATION EXCHANGE. The replacement of the cations adsorbed on colloidal particles by other cations. Also termed *base exchange*.

CATION EXCHANGE CAPACITY. A measure of the ability of a colloidal particle to adsorb cations. It is expressed in milliequivalents of the adsorbed cation per 100 grams of the colloid.

CHARGE-DENSITY. The ratio of the charge of a colloidal particle expressed in valence units or cation exchange capacity, to the surface area in square meters.

CHELATION. The reaction between a metallic ion and a complexing agent, generally organic, with the formation of a ring structure and the effective removal of the metallic ion from the system.

CLEAVAGE. The tendency of a crystalline substance to fracture in certain definite directions yielding smooth surfaces.

COLLOID. Finely divided particles with diameters in the range of 10^{-5} to 10^{-7} cm.

CONCENTRATION RATIO. The ratio of the content of a constituent at a specific level within a weathered sequence to its content in the parent material.

CO-ORDINATION NUMBER (OF A CATION). The number of anions that may surround and bond to a cation.

COVALENT BOND. A linkage between two atoms in a molecule by the sharing of electron pairs.

CRYPTOCRYSTALLINE. Crystalline particles that are too fine to be resolved by normal optical means.

CRYSTALLOGRAPHIC AXIS. An imaginary line used as a co-ordinate axis of reference in symbolizing planes within a crystal. The axes are designated a, b, and c.

DESILICIFICATION (or Desilication). The loss of silica through solution.

DIABASE. An intrusive basaltic rock.

DIAGENESIS. Physical and chemical changes occurring in sediments during and after deposition and burial, including compaction, consolidation, cementation and authigenesis.

DIKE. A tabular body of igneous rock that intersects larger igneous masses or cuts across the bedding planes of sedimentary rocks.

DOLERITE. See DIABASE.

DYKE. See DIKE.

Eh. See REDOX POTENTIAL.

ELECTRODIALYSIS. The separation of ions in solution from colloidal particles by means of an electric potential and a semi-permeable membrane.

ELECTRONEGATIVITY. A measure of the tendency of an element to form negative ions. (Also the strength of the ionic bond of a cation.)

ELUVIATION. The downward or sideward movement and subsequent precipitation of soluble or suspended material within a soil profile.

ENDOTHERMIC REACTION. A reaction which involves an energy change manifested by the absorption of heat.

EPEIROGENY. Broad uplifts or downwarps of continental masses or ocean floors.

EXOTHERMIC REACTION. A reaction which involves an energy change manifested by the evolution of heat.

FIXATION. The absorption of ions by crystalline particles or organic matter so that the ions cannot be recovered by simple exchange reactions.

FLOCCULATION. The aggregation of colloids to form floccules or relatively coarse particles.

GEODE. A subspherical cavity in a rock generally lined with crystals.

GLEI. A blue-gray soil zone of permanent water saturation occurring toward the base of certain soil profiles.

GLEY. See GLEI.

GNEISS. A banded coarse-grained metamorphic rock.

GRANITE. A coarse-grained acid igneous rock composed essentially of alkali feldspar and quartz with various accessory minerals.

HALMYROLOSIS. Submarine weathering of rocks and sediments.

HARDPAN. A hard, impervious clay layer occurring in certain soils.

HEAVY MINERAL. A chemically stable, accessory mineral with a specific gravity greater than that of bromoform (2.89), occuring in sedimentary rocks.

HORIZON (SOIL). A clearly defined layer within a soil profile.

HORNFELS. A massive fine-grained metamorphic rock formed under the influence of a high temperature.

HUMID CLIMATE. One in which precipitation exceeds evaporation.

HYDRATION. The combination of an element or compound with water.

HYDROGEN BOND. A weak type of bonding formed betewen two adjacent oxygens by an oscillating hydrogen ion.

HYDROLYSIS. The chemical decomposition of a substance by water, the water itself being also decomposed.

HYDROTHERMAL ACTIVITY. The alteration of rocks and minerals by the action of heated or superheated aqueous solutions.

IGNEOUS ROCK. Rocks formed by solidification from a molten or partly molten state. Such rocks are termed ACID where the silica content exceeds 66% and BASIC where the value lies between 45% and 54%.

ILLUVIATION. The deposition within a soil of soluble or suspended material that has been derived from a higher level in the solum.

INTERSTRATIFICATION (OF CLAY MINERALS). Refers to the presence of two or more layered clay minerals in a single crystal. Also termed *mixed-layering*.

IONIC BOND. The linkage between two atoms involving the transfer of electrons from one to the other and resulting in the development of opposite charges on the two atoms.

IONIC POTENTIAL. The ratio of the charge of an ion expressed in valence units to its radius in angstrom units.

ISOMORPHOUS SUBSTITUTION. The substitution of one element for another within a crystal structure without appreciable change to that structure.

ISOSTRUCTURAL. Minerals of different chemical composition, that possess similar crystal structures.

LATERITE. A weathered residue composed essentially of ferric oxide and hydroxide minerals such as hematite, maghemite and goethite.

LOESS. Loosely consolidated silt, occasionally with fine sand and clay, deposited by wind.

MAGMATIC ROCKS. See IGNEOUS ROCKS.

METAMORPHISM (OF ROCKS). The process by which consolidated rocks are altered in composition, texture and structure by pressure, heat or chemically active fluids.

METEORIC WATER. Water of atmospheric origin.

MIXED LAYERS. See INTERSTRATIFICATION.

OXIDATION. The increase in the positive valence or the decrease in the negative valence of an element.

PARENT MATERIAL. The rock or minerals from which the weathered material has been derived.

PEDALFERS. Acid soils of the humid regions containing a zone of sesquioxide $(Fe_2O_3 + Al_2O_3)$ accumulation in the solum.

PEDOCALS. Alkaline soils generally of arid or semi-arid regions, containing a zone of carbonate accumulation in the solum.

PEGMATITE. Very coarse-grained dike rocks generally occurring within massive igneous rocks.

PENEPLAIN. Almost a plain resulting from the extensive erosion of a land surface.

PERIDOTITE. An ultrabasic igneous rock rich in olivine.

pH. The degree of alkalinity of acidity of a system expressed as the logarithm to the base 10 of the reciprocal of the hydrogen ion concentration in grams per litre.

PISOLITE. A spherical or subspherical mass of fine-grained crystalline or non-crystalline particles arranged in a concentric structure.

PLAYA. A temporary or ephemeral lake in arid regions.

POLARIZATION (OF IONS). The distortion of the electron shells from symmetrical distribution about the nucleus of an ion.

POLAR MOLECULES. Molecules which possess a dipole moment, that is, are bonded in such a manner that one end of the bond is more positive than the other.

POLYCRYSTALLINE. An aggregate of crystalline particles.

POLYMERIZATION. A reaction between two or more molecules of the same compound to form larger molecules with the same empirical formula but with a multiple of the molecular weight.

POLYMORPHISM. A chemical compound that crystallizes with two or more distinct crystal structures.

PORPHYRY. An igneous rock composed of relatively coarse-grained crystals set in a fine-grained groundmass.

PSEUDOMORPH. The replacement of one mineral by another so that the second mineral assumes the crystal shape or form of the first.

PYROCLASTIC ROCK. Rock material that has been explosively or aerially ejected from a volcanic vent.

REDOX POTENTIAL (Eh) (or Oxidation–Reduction potential). A measure of the ability of a system to bring about an oxidation or reduction reaction.

REDUCTION. Is the increase in the negative valence or the decrease in the positive valence of an element.

RHYOLITE. A fine-grained, silica-rich, volcanic rock.

SCHIST. A medium-grained, metamorphic rock possessing a sub-parallel orientation of flakey minerals such as chlorites, micas, and amphiboles.

SESQUIOXIDES. Oxides of aluminum and ferric iron.

SILICIFICATION. The addition of silica generally but not always by precipitation from solution.

SOLUBILITY PRODUCT. The product of the concentration of the constituent ions of a compound in water when in equilibrium with the non ionized compound.

SOLUM. The upper part of a soil profile and includes the A and B horizons.

SPHERULITES. Spherical aggregates of crystalline or non crystalline particles arranged in radiating structures.

SYENITE. A coarse-grained igneous rock composed principally of alkali feldspar generally in association with amphiboles, pyroxenes or biotite.

TRACHYTE. A volcanic rock composed essentially of alkali feldspar with or without amphiboles, pyroxenes or biotite.

TRACHYANDESITE. A volcanic rock rich in plagioclase but containing some alkali feldspar usually in association with hornblende and biotite.

UNIT CELL. The smallest part of a crystal or crystalline mass that contains the molecular structure and is repetitive throughout the crystal or mass.

VESICLE. A small subspherical cavity in a volcanic rock formed by the expansion of a gas or steam bubble during solidification of the rock.

WATER-TABLE. The upper limit of the zone of water saturation within the earth's crust.

X-AMORPHOUS. Substances which fail to show a crystalline structure upon examination with X-ray diffraction. Such substances may possess a crystalline structure that is detectable by electron diffraction or other means.

Author Index

Numbers set in *italics* designate the page numbers on which the complete literature citations are given.

143

Subject Index